AF302468

Current Topics in Microbiology 143 and Immunology

Editors

R. W. Compans, Birmingham/Alabama · M. Cooper, Birmingham/Alabama · H. Koprowski, Philadelphia
I. McConell, Edinburgh · F. Melchers, Basel
V. Nussenzweig, New York · M. Oldstone, La Jolla/California · S. Olsnes, Oslo · H. Saedler, Cologne · P. K. Vogt, Los Angeles · H. Wagner, Ulm
I. Wilson, La Jolla/California

In Situ Hybridization

Edited by
A. T. Haase and M. B. A. Oldstone

With 33 Figures, 8 in colored

Springer-Verlag
Berlin Heidelberg New York
London Paris Tokyo Hong Kong

Ashley T. Haase, M. D.

Department of Microbiology
University of Minnesota Medical Scool
Minneapolis, MN 55455, USA

Michael B. A. Oldstone, M. D.

Department of Immunology
Scripps Clinic and Research Foundation
10666 North Torrey Pines Rd., La Jolla, CA 92037, USA

ISBN-13:978-3-642-74427-3   e-ISBN-13:978-3-642-74425-9
DOI: 10.1007/978-3-642-74425-9

This work is subject to copyright. All rights are reserved, whether the whole or part of the material is concerned, specifically the rights of translation, reprinting, reuse of illustrations, recitation, broadcasting, reproduction on microfilms or in other ways, and storage in data banks. Duplication of this publication or parts thereof is only permitted under the provisions of the German Copyright Law of September 9, 1965, in its version of June 24, 1985, and a copyright fee must always be paid. Violations fall under the prosecution act of the German Copyright Law.

© Springer-Verlag Berlin Heidelberg 1989
Softcover reprint of the hardcover 1st edition 1989

Library of Congress Catalog Card Number 15-12910

The use of registered names, trademarks, etc. in this publication does not imply, even in the absence of a specific statement, that such names are exempt from the relevant protective laws and regulations and therefore free for general use.

Product Liability: The publisher can give no guarantee for information about drug dosage and application thereof contained on this book. In every individual case the respective user must check its accuracy by consulting other pharmaceutical literature.

2123/3020-543210 — Printed on acid-free paper

Foreword

The ability to locate viral genes, transcripts and proteins to unique cells and tissues in vivo while preserving morphology has been a powerful tool for understanding pathogenesis. Because of the several ingenious ways in situ hybridization can be used, and in consideration of some difficulties with the technology, this volume was conceived to address both of these issues. Ashley Haase, who has been one of the leaders in developing and popularizing the utility of this approach for the study of pathogenesis, has admirably put together this volume with these views in mind.

Michael B. A. Oldstone
Research Institute of Scripps Clinic

Preface

Viral infections can be conceptually characterized as a series of discrete events, beginning with invasion of the host and continuing with multiplication at the site of entry and subsequent dissemination throughout the organism. The fate of the virus (or the host, depending on one's viewpoint), is determined by the nature of the interactions of virus and host cells and by the general and specific immune defenses mounted in response to infection. Fortunately, of course, these measures usually prevail, and the offending agent is cleared with little more harm than a brief self-limited illness largely reflecting damage to the cells and tissues involved in the initial stages of infection. Far less frequently, these destructive consequences of viral replication are so extensive that serious illness or death is the result; or, at the other end of the scale, the virus is not eliminated and a persistent infection ensues which may eventually cause disease many years later. Human immunodeficiency virus infection and AIDS is the most prevalent and alarming contemporary paradigm of these slow infections caused by viruses and virus-like agents.

Several fundamental and central issues follow logically from this way of looking at the pathogenesis of infections. Which cells serve as hosts in the initial stages of infection and in the ensuing spread and replication at secondary sites? Will the host cell fully or only partially support the growth of virus? What is to be the fate of the infected cell — death, dysfunction, transformation, or no apparent ill effect at all? What is the outcome of the clash between viruses and host defenses — eradiction or persistence?

These questions ultimately must be answered within the complex setting of the whole organism. It is for this reason that virologists have adapted or invented increasingly sophisticated methods to grapple with these issues in relevant contexts. Understanding the mechanisms of infection rests on insights derived originally from quantitative assays of infectious virus in organ systems and light and electron microscopic examination of the tissues to visualize viral particles and the pathological sequelae of replication. These methods have complemented rather than supplanted immunofluorescence and immunocytochemistry to locate viral gene products in cells, thus pushing the analyses one step earlier in the viral life cycle.

The most recent and important technological development in this series is in situ hybridization to detect, quantitate, and locate viral genes in particular cell types in defined anatomical sites. Because only viral genomes and transcripts at low abundancy need be present, in situ hybridization has opened new windows onto the world of covert infections in which viral gene expression is so restricted that a search for the usual indications of infection (tissue damage, viral particles and proteins, or infectious virions) will likely prove unrewarding. Because only a small fraction (generally less than one in a hundred to a thousand) of the cells in tissues harbor viruses, in situ hybridization is particularly useful, if not absolutely essential, in analyses of infections on the organismal level, where other methods of evaluation involving extraction of tissues will dilute out the virus-specific components beyond detection, as well as sacrificing valuable anatomical information.

In in situ hybridization, preparations of cells or tissue sections are annealed to a virus-specific probe with a radioactive or nonradioactive reporter group. Methods which retain cellular morphology and the viral nucleic acid target with sensitivities in the range of a few copies of viral genomes or mRNAs are reviewed in Haase (1986, 1987). These publications also contain descriptions of methods, controls for specificity, and ample documentation of the utility of in situ hybridization in deepening our understanding of virus host cell tropisms, dissemination of viruses, and the role of restricted gene expression in persistent, covert and slow infections (Haase et al. 1977, 1981, 1984, 1985a, b, 1986; Peluso et al. 1985).

This volume represents the advancing salients of in situ hybridization technology. Jack Stevens describes the convergences of recombinant DNA technology which generates single-stranded probes from different regions of the herpes simplex virus (HSV) genome, with extant in situ hybridization technologies, defining (and revealing) the extraordinary specificity and restriction of viral gene expression in latency. Brahic and Haase in another convergence, have devised double-label methods for analyses of mechanisms of viral persistence in vivo which combine immunocytochemistry, photographic chemistry and in situ hybridization to unequivocally identify the type of cells that serve as hosts for viruses, and to measure both viral genes and their products. In their chapter of the automation of in situ hybridization, Unger and Brigati extend the nonisotopic approaches to in situ hybridization, with all their inherent advantage of speed and safety, to the diagnostic setting. Lipkin, Villarreal and Oldstone have also extended to the whole organism the technology of methods which have already served so well in their studies of viral tropisms, changes in tissue distribution and clearance of viral genes and products in the course of infection, and expression of specific genes of virus and host in natural infection or with experimental manipulation. Finally, at the other end of the anatomical scale, Singer takes in situ hybridization to the subcellular level, with newer methods and probes which provide high resolution electron micrographs locating specific transcripts. The editors believe these advances collectively offer new opportunities to understand viral infections at the molecular level.

Ashley T. Haase

References

Haase AT (1986) Analysis of viral infections by in situ hybridization. J Histochem Cytochem 34: 27—32

Haase AT (1987) Analysis of viral infections by in situ hybridization. In: Valentino K, Roberts, J, Barchas J (eds) In situ hybridization: Applications to neurobiology. Oxford University Press, New York, pp. 197–219

Haase AT, Stowring LS, Narayan O, Griffin D, Price D (1977) Slow persistent infection caused by visna virus: role of host restriction. Science 195: 175–177

Haase AT, Ventura P, Gibbs CJ, Tourtellotte WW (1981) Measles virus nucleotide sequences: detection by hybridization in situ. Science 212: 672–675

Haase AT, Brahic M, Stowring L (1984) Detection of viral nucleic acids by in situ hybridization. In: Maramorosch K, Koprowski H (eds) Methods in virology, vol 7 Academic, New York, pp 189–226

Haase AT, Gantz D, Eble B, Walker D, Stowring L, Ventura P, Blum H, Wietegrefe S, Zupanic M, Tourtellotte W, Gibbs CJJr, Norrby E, Rozenblatt S (1985a) Natural history of restricted synthesis and expression of measles virus genes in subacute sclerosing panencephalitis. Proc Natl Acad Sci USA 82: 3020–3024

Haase AT, Stowring L, Ventura P, Burks J, Ebers G, Tourtellotte W, Warren K (1985b) Detection by hybridization of viral infection of the human central nervous system. Ann NY Acad Sci 436: 103–108

Haase AT In situ hybridization and covert virus infections (1988) Notkins AL, Oldstone MBA (eds) In: Concepts in viral pathogenesis, vol 2. Springer, New York pp. 310–316

Peluso R, Haase AT, Stowring L, Edwards M, Ventura P (1985) A Trojan horse mechanism for the spread of visna virus in monocytes. Virology 147: 231–236

Table of Contents

Indexed in Current Contents

List of Contributors

You will find the addresses at the beginning of the respective
contribution

J. B. LAWRENCE
S. BILLINGS-GAGLIARDI
M. BRAHIC
D. J. BRIGATI
A. T. HAASE
G. L. LANGEVIN
W. I. LIPKIN

M. B. A. OLDSTONE
M. POMEROY
F. SILVA
R. H. SINGER
J. G. STEVENS
E. R. UNGER
L. P. VILLARREAL

Herpes Simplex Virus Latency Analyzed by In Situ Hybridization*

J. G. STEVENS

1 Introduction

To date, in situ nucleic acid hybridization technology has proved singularly useful in generating information about the basic properties of latent herpes simplex virus (HSV) infections in vivo. In adapting extant methods to detect viral transcripts, convincing data have been gathered which identify (a) the cell in which the latent infection is established, and (b) the nature of viral transcripts present in latently infected cells. These ongoing descriptive studies provide the necessary insight and theoretical framework for investigations into the underlying mechanisms involved in latent infection. In addition to studies of latently infected cells, in situ methodology has been employed in several laboratories to investigate the pathogenesis of acute infection, largely confirming the conclusions earlier inferred from classical methods. This chapter focuses on the recent successes in analyses of latency made possible by the convergence of recombinant DNA technology, which generates single stranded probes of defined polarity and physical location on the viral genome, and increasingly sensitive techniques for in situ hybridization.

2 Identification of Latently Infected Cells

The demonstration that herpes simplex virus establishes latent infections in sensory ganglia and the central nervous system (STEVENS and COOK 1971, KNOTIS et al. 1973, reviewed in STEVENS 1975) stimulated efforts to determine which cell type harbors latent viral genomes. Direct approaches to this problem can

Department of Microbiology and Immunology, UCLA School of Medicine, Los Angeles, California 90024-1747, USA

* The research from the author's Laboratory which is discussed here was supported by grant AI-06246 from the National Institute of Health, and RG 1647-AI from the National Multiple Sclerosis Society.

readily be envisioned — for example, by physically separating cell types populating spinal ganglia or the central nervous system, co-cultivating the different classes of cells with susceptible indicator cells, scoring reactivation by assessing viral cyto-pathology on the indicator cells, and confirming identity of the reactivated virus by neutralization with a specific antiserum. However, to achieve the separations without also destroying the cells to be employed in reactivation studies were and still are unavailable. Therefore, more indirect approaches employing histological techniques have been used on ganglia, in which latent virus was reactivated by the "stresses" of in vitro ganglionic cultivation (see STEVENS and COOK 1973; STEVENS 1975). In these studies, the purpose was to define the initial cells in which viral antigens, viral nucleic acids, or morphologically identifiable viral components could be detected during the reactivation process. The assumption was made that this cell harbored the viral genetic information. As is discussed further below, in situ hybridi-zation technology was one of the most useful procedures employed in these early studies, and by all the methods used, neurons were identified as the cell which perpetuates infection. In attempts to rule out passage of virus from supporting cells to neurons with amplification of viral gene products only in the latter as an explanation for these results, McLENNON and DARBY (1980) used a temperature-sensitive mutant of HSV-1 which established a latent infection in murine spinal ganglia. It was hypothesized that when reactivation was induced by in vitro cultivation of ganglia at the restrictive temperature, antigens would appear only in cells harboring latent viral DNA (abortively infected because the virus is temperature sensitive). In these experiments, antigens did in fact appear only in neurons, providing additional evidence that the latent viral genome is sequestered in neurons.

These relatively primitive studies of 15 years ago, utilizing in situ hybridization, have recently been confirmed by more sophisticated technologies which combine molecularly cloned probes and in situ hybridization to identify specific viral transcripts in latently infected neurons. In our early studies, c-RNA probes made by transcribing viral DNA with *Escherichia coli* RNA polymerase were employed. In agreement with the other results described above, viral DNA was intially detectable in neurons and *then* passed (in some form) to supporting cells (COOK et al. 1974). It is important to note here that viral DNA was not detected in latently infected neurons not induced to reactivate, even though reactivating neurons were easy to detect by these methods. This latter result has been a general one in all laboratories using in situ technology involving RNA and DNA probes to detect HSV-1 DNA in latently infected cells, and, to my knowledge, no one has been able to reproducibly detect the latent viral DNA in these cells. Although an occasional neuron can be found to be expressing easily detectable viral DNA at the time of explant, our experience is that the number of such cells is increased some 100-fold after 3 days of ganglionic cultivation in vitro. We interpret the rare neuron found before cultivation to be spontaneously reactivating, and the vast majority of latently infected cells to be harboring DNA at a level too low to be detectable by the methods employed (see STEVENS 1975 for a more detailed discussion of this subject). This inability to detect latent viral DNA is unexpected, since DNA genomes con-siderably less complex than HSV DNA are present at levels of approximately one copy per latent cell and are detectable using similar methods in other viral systems (cf. HAASE et al. 1982; BLUM et al. 1983) and may imply sequestration of DNA in impermeable structures.

Definitive identification of latently infected cells by nucleic acid hybridization technologies has been accomplished more recently in experiments directed to the nature of viral transcripts in latency. Viral RNA was detected in sensory neurons in ganglia derived from humans (GALLOWAY et al. 1982) or guinea pigs (TENSER et al. 1982) latently infected with HSV-2 with nick-translated, double-stranded DNA probes. In the first experiments with HSV-1, STROOP et al. (1984) inoculated mice on the corneas with virus and examined the trigeminal ganglia and the central nervous system of survivors with nick-translated, ^{3}H-labeled full-length genomic probes isolated from virions. DNA and RNA were both found in the acute phase, but only RNA was detectable during the latent phase of infection. During the latent phase, significantly increased numbers of grains in the radioautograph were detectable only over neurons, predominantly over the nuclei. Subsequently, using molecularly cloned (a) double-stranded, ^{3}H- or ^{35}S-labeled, nick-translated DNA probes (STEVENS et al. 1987, ROCK et al. 1987; DEATLY et al. 1988; WAGNER et al. 1988), (b) single-stranded, ^{3}H-labeled DNA probes prepared by primer extension (STEVENS et al. 1988), and (c) ^{35}S- and ^{3}H-labeled RNA probes prepared by primer extension (CROEN et al. 1987; ROCK et al. 1987; GORDON et al. 1988), neurons in sensory ganglia from mice, rabbits, and humans were selectively labeled when probed for RNA using in situ methods. In all cases, the majority of grains were found to be nuclear. Although some cells identified as glial and satellite cells were found by DEATLY et al. (1988) to express viral RNA, these were small in number compared to the number of neurons, and detection of these cells has not been a finding generally reported. To summarize this section the described studies collectively provide a strong argument that neurons harbor latent HSV-1.

3 Characterization of HSV-1 Transcripts in Neurons

In the earlier studies concerned with a precise identification of transcripts present in neurons latently infected with HSV-1, our group probed latently infected murine spinal ganglia with four pools of molecularly cloned probes representing the entire HSV-1 genome (STEVENS et al. 1987). In these experiments, the ganglia were cut on a cryostat and fixed with ethanol acetic acid (3:1), on slides which had previously been treated with DENHART's solution and acetylated for 15 min at room temperature (HAASE et al. 1984). The probes were nick-translated to specific activities of greater than 10^7 cpm/µg with ^{3}H-labeled nucleotides. The hybridization was carried out as described by HAASE et al. (1984), with the conditions of probe "prehybridization" and increased stringency during hybridization and washing taken from STROOP et al. (1984). Conditions for deoxyribonuclease treatment of slides scored for RNA were also those described by HAASE. Although in our experience the high stringency wash which STROOP et al. described is particularly important, this can be modified somewhat by first using "low stringency conditions" [50% formamide, 10 mM TRIS pH 7.4, 1 mM ethylene diaminetetraacetic acid (EDTA) pH 8.0, 0.6 mM NaCl] for 72 h at room temperature with frequent changes, followed by a 2-h "high stringency" wash (50% formamide, 10 mM TRIS pH 7.4, 1 mM EDTA pH 8.0, 2 × SSC) at 45 °C. These procedures assure reproducibly low background counts. Using the conditions of nick translation described, probes approximately 40 nucleotides in length are generated. In our hands, these relatively short molecules are required for maximal sensitivity for the method.

With this procedure, the mixed probes prepared from all areas of the genome were found to hybridize to acutely infected tissues, but only those representing the repeat regions of the viral genome hybridized to latently infected neurons. As the figures show, with the high resolution of ³H-labeled probes, transcripts in latently infected neurons were restricted to nuclei (Fig. 1 a), whereas in acutely infected cells probed with the same DNA (Fig. 1 b), transcripts were detected in cytoplasm as well. In additional experiments, principally involving Northern blotting technologies, we showed (1987) that the in situ signal was arising from a unique and

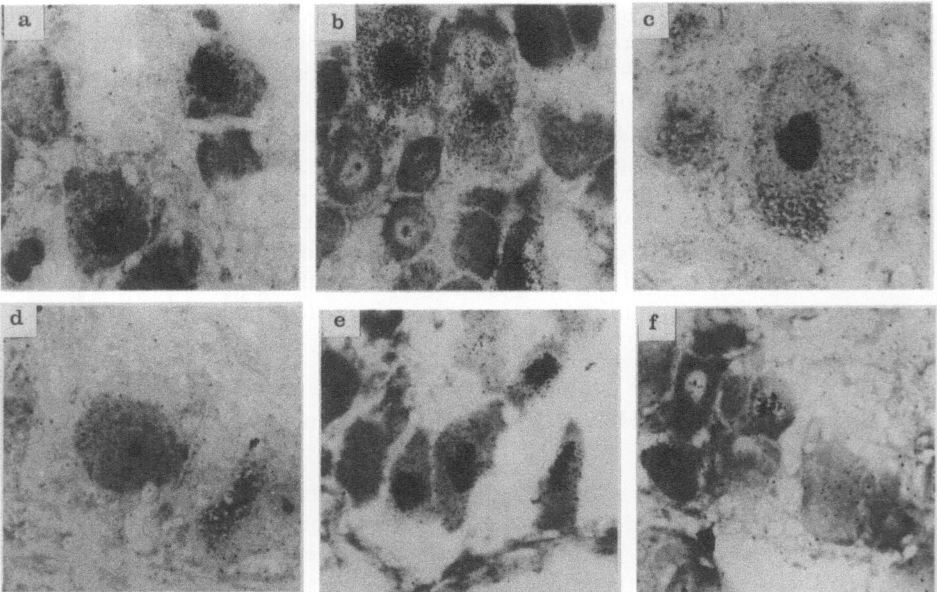

Fig. 1 a–f. In situ nucleic acid hybridization of neurons in sensory ganglia with ³H-labeled HSV-1 DNA probes. **a** Latently infected spinal ganglion with a double-stranded *Sal*I-*Bam*HI fragment (0.79–0.81 map units) specific for LAT or ICPO. Silver grains are restricted to the nucleus, and only LAT is detected. **b** Acutely infected murine spinal ganglion, same probe as **a**. Silver grains over nucleus and cytoplasm, detecting both ICPO and LAT. **c** Latently infected human trigeminal ganglion hybridized with a single-stranded ³H-labeled *Sal*I-*Bcl*I fragment (0.790–0.798 map units) specific for LAT. Silver grains are concentrated over the nucleus. **d** Trigeminal ganglia from an HSV seropositive individual hybridized to probe employed in **d**. Background labeling. **e** Latently infected murine spinal ganglia hybridized with a double-stranded *Pst*I-*Hpa*I fragment (0.775–0.785 map units) specific for LAT. Silver grains are restricted to the nucleus. **f** Latently infected murine spinal ganglion hybridized with a double-stranded *Bam*HI-*Sst*I fragment (0.807–0.815 map units) detecting a transcript 3′ to LAT, and associated with latent infection. Concentrations of silver grains are considerably reduced compared to **a** and **e**. All magnifications × 300. Map positions of the fragments used as probes are graphically presented in Fig. 3

previously undescribed HSV transcript in the long terminal repeat region complementary to the gene encoding the immediate early protein ICPO. This transcript was termed "latency associated transcript" or LAT. Thus, this combination of hybridization procedures defined a unique transcription pattern for HSV-1 in latently infected neurons.

Extension of these fundamental observations to further characterize latency has relied heavily on in situ hybridization technology. It was shown by ROCK et al. (1987) (using smaller fragments covering the ICPO gene region) that a similar transcript, and an additional one present more rarely, could be detected in latently infected rabbits. It was also suggested that these transcripts may well map at opposite ends of the ICPO gene. In addition, hybridization experiments with individual *EcoRI* fragments representing the entire genome (13 fragments supplemented with *Bam*HI fragments appropriate for crucial areas), when employed individually, indicated that only the areas represented by long terminal repeats were transcribed. DEATLY et al. (1988) showed that transcripts from similar areas were present in latently infected murine CNS.

Comparable studies in human trigeminal ganglia by three groups were completely concordant (CROEN et al. 1987; GORDON et al. 1988; STEVENS et al. 1988). No transcripts from other areas of the genome were detected when the probes depicted in Fig. 2 were hybridized to latently infected human ganglia. It is of interest to point out that human tissues taken from cadavers as long as 60 hours after death demonstrated positive signals with probes employed. Clearly, sufficient polymerized RNA was present to insure nucleation of probe sequences under these conditions, and the result is somewhat surprising in view of the drastic effects which ribonuclease is generally considered to have under these kinds of conditions. An example of the

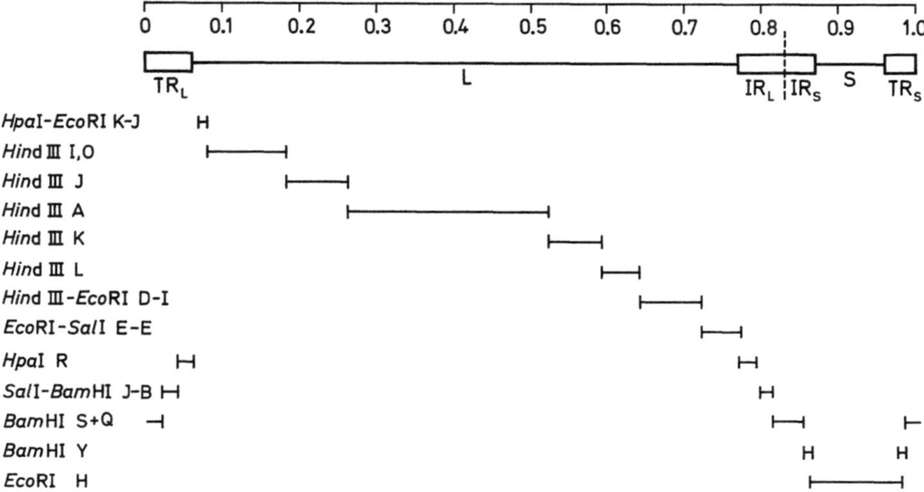

Fig. 2. HSV-1 DNA fragments used as probes to detect transcripts in human trigeminal ganglia. Fragments detecting signal in latently infected neurons were *Hpa*IR, *Sal*I-*Bam*HI JB and *Bam*I S + Q

type of signal one gets on these tissues with a single-stranded DNA probe specific for LAT is shown in Fig. 1 c.

Although quantitative aspects and sensitivities of the technology have not been completely established, we estimate that there are more than 10^4 molecules of LAT in latently infected neurons of mice. In humans it was shown that when the *Sal Bam* J-B fragment was added to a mixture of all the genomic probes which failed to generate a signal (i.e., all but *Sal*I *Bam*HI J-B, *Hpa*I R, and *Bam*HIS + Q, see Fig. 2), LAT was easily detectable by our standard *in situ* hybridization method. Since the *Sal Bam* J-B fragment represents approximately 1/70 of the viral genome, it was concluded that if another gene were expressed at levels equivalent to the LAT transcript detectable with the probe present, we should have found it (STEVENS et al. 1988). Actually, senstivity is likely to be better than this, since the *Sal*I *Bam*HI J-B fragment covers only about 1/3 of the LAT transcript.

It has not been possible as yet to precisely map and characterize the all LAT transcripts by S1 nuclease, primer extension and sequencing techniques, because of the relatively low level of LAT in tissues. Nonetheless, in situ hybridization with additional probes from the region in question has defined one major transcript with its 5' terminus downstream from the termination of the ICPO gene, and a 3' terminus within the ICPO gene (WAGNER et al. 1988). An example of the kind of experiment which places the 5' end of LAT downstream from the ICPO gene is shown in Fig. 1e. Here, a probe 3' to the terminus of ICPO was employed in the in situ hybridization. In addition, as mentioned earlier, a "minor" RNA related to the 5' end of ICPO was detected intially by ROCK et al. (1987). Through the use of strand specific probes, we found this transcript or transcripts to be transcribed from the same DNA strand as LAT, and present in about 1/10 the neurons expressing LAT. In addition, the number of silver grains per neuron (Fig. 1f) was always significantly less than that obtained with probes for LAT used on the same ganglion, suggesting that the number of transcripts may be reduced relative to LAT. The transcript was also transcribed from the same strand as LAT, and co-expressed with that molecule (J. G. STEVENS et al unpublished). In these latter experiments, a mutant lacking the regulatory region and approximately 1/4 of the 5' transcribed sequences

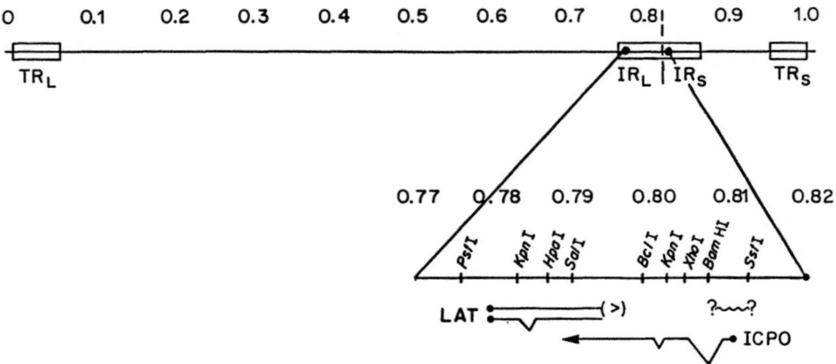

Fig. 3. Map positions of "major" and "minor" LAT and the ICPO message. The restriction endonuclease sites shown identify fragments employed for probes in Fig. 1.

of LAT was tested and found to establish latent infections from which it could be reactivated by in vitro cultivation of murine spinal ganglia. When these ganglia harboring latent virus were probed by in situ methods for viral transcripts (using probes covering the entire viral genome), *none* were found. Although the purpose here is to indicate that expression of the minor transcript is linked to the expression of LAT, the more significant general observation is that latent HSV-1 infections can be maintained in mice in the absence of *any* transcripts detectable by our methods of in situ hybridization (JAVIER et al. 1988).

Finally, using northern blotting methods, it has been shown (SPIVAK and FRASER 1987; WAGNER et al. 1988) that at least one and possibly two additional transcripts are present in these neurons, that the sequences are related, and that one of the additional transcripts is a spliced derivative of the larger transcript originally described (WAGNER et al., 1988). A summary of our view of the present state of knowledge of LAT is presented in Fig. 3. Although two principal molecules are shown (plus the minor one related to the 5′ terminus of the ICPO gene), one additional molecule related to the major transcript has been described by SPIVAK and FRASER (1987). This figure also depicts the map position of fragments used for probes employed in the in situ hybridization experiments in Fig. 1.

4 Conclusions

From the results discussed here, it is clear that in situ hybridization technology has played a major role in providing insight into the basic properties of herpes simplex virus latency. These studies in fact represent one of the best examples of application of this technology to a problem which cannot be adequately studied by conventional biochemical or molecular biological procedures because sensory ganglia do not represent pure cultures of latently infected neurons (at most, these cells constitute 5% of the ganglionic cell population). Systematic and sustained application of these powerful in situ technologies to the herpesviruses in which the latent state remains largely undefined (most notably cytomegalovirus and varicella-zoster virus) should yield important imformation and concepts. For herpes simplex, in situ hybridization methods, coupled with the finely mapped probes, provide an opportunity to understand viral gene expression accompanying the reactivation process.

References

Cook ML, Bastone VB, Stevens JG (1974) Evidence that neurons harbor latent herpes simplex virus. Infec Immun 9: 946–951

Croen KD, Ostrove JM, Dragovic LJ, Smialek JE, Strauss SE (1987) Latent herpes simplex virus in human trigeminal ganglia: detection of an immediate early gene "anti-sense" transcript by in situ hybridization. New Eng J Med 317: 1427–1432

Deatly AM, Spivak JG, Lavi E, O'Boyle II D, Fraser NW (1988) Latent herpes simplex virus type 1 transcripts in peripheral and central nervous system tissues of mice map to similar regions of the viral genome. J Virol 62: 749–756

Galloway DA, Fenoglio CM, McDougall JK (1982) Limited transcription of the herpes simplex virus genome when latent in human sensory ganglia. J Virol 41: 686–691

Gordon YJ, Johnson B, Romanowski E, Aravello-Cruz T (1988) RNA complementary to herpes simplex virus type 1 ICPO gene demonstrated in neurons of human trigeminal ganglia. J Virol 62: 1832–1835

Haase AT, Stowring L, Harris JD, Traynor B, Venturi P, Peluso R, Brahic M (1982) Visna DNA synthesis and the tempo of infection in vitro. Virology 119: 399–410

Blum HE, Stowring L, Figus A, Montgomery CK, Haase AT, Vyas GN (1983) Detection of hepatitis B virus DNA in hepatocytes, bile duct epithelium, and vascular elements by in situ hybridization. Proc Natl Acad Sci USA 80: 6605–6688

Haase A, Brahic M, Stowring L and Blum H (1984) Detection of vira/nucleic acids by *in situ* hybridization. Methods Virol 7: 189–226

Javier RT, Stevens JG, Dissette VB, Wagner EK (1988) A Herpes simplex virus transcript abundant in latently infected neurons is dispensable for establishment of the latent state. Virology 166: 254–257

Knotts FB, Cook ML and Stevens JG (1973) Latent herpes simplex virus in the central nervous system of mice and rabbits. J Exptl Med 138: 740–744

McClennon JL and Darby G (1980) Herpes simplex virus latency: the cellular location of virus in dorsal root ganglia and the fate of the infected cell following virus activation. J Gen Virol 51: 233–243

Rock DL, Nesburn AB, Ghiasi H, Ong J, Lewis TL, Lokensgard JR and Wechsler SL (1987) Detection of latency-related viral RNAs in trigeminal ganglia of rabbits latently infected with herpes simplex virus type 1. J Virol 61: 3028–3826

Spivak JG and Fraser NW (1987) Detection of herpes simplex virus type 1 transcripts during latent infection in mice. J Virol 61: 3841–3847

Stevens JG (1975) Latent characteristics of selected herpesviruses. Adv Career Res 26: 227–256

Stevens JG and Cook ML (1971) Latent herpes simplex virus in spinal ganglia of mice. Science 173: 843–845

Stevens JG, Cook ML (1973) Latent herpes simplex virus in sensory ganglia. In: Perspectives in Virology III. Academic, New York

Stevens JG, Wagner EK, Devi-Rao GB, Cook ML, Feldman LT (1987) RNA complementary to a herpesvirus α gene is prominent in latently infected neurons. Science 235: 1056–1059

Stevens JG, Haarr L, Porter DD, Cook ML, Wagner EK (1988) The herpes simplex virus latency-associated transcript is prominent in trigeminal ganglia from seropositive humans. J Inf Dis 158: 117–123

Stroop WG, Rock DL, Fraser NW (1984) Localization of herpes simplex virus in the trigeminal and olfactory systems of the mouse central nervous system during acute and latent infections by in situ hybridization. Lab Invest 51: 27–38

Tenser RB, Ressel SJ, Dunston ME (1982) Detection of herpes simplex virus mRNA in latently infected trigeminal ganglion neurons by in situ hybridization. Ann Neurol 11: 285–291

Wagner EK, Devi-Rao GB, Feldman LT, Dobson AT, Zhang Y-F, Flanagan WM, Stevens JG (1988) Physical characterization of the herpes simplex virus latency-associated transcript in neurons. J Virol 62: 1194–1202

Wagner EK, Flanagan WM, Devi-Rao G, Zhang Y-F, Hill JM, Anderson KP, Stevens JG (1988) The Herpes simplex virus latency associated transcript is spliced during the latent phase of infection. J Virol 62: 4517–4585

Double-Label Techniques of in Situ Hybridization and Immunocytochemistry*

M. Brahic[1] and A. T. Haase[2]

1 Introduction

The outcome of viral infection is largely determined by the types of cells the virus invades and the steps of viral replication which the cell will support. Under permissive conditions, high levels of viral gene expression in cells essential to the host may result in serious illness or death. On the other hand, persistent infections are the result of restricted gene expression which may allow the host cell to survive and the virus to escape detection and destruction by the host's immune surveillance system. We have devised two double-label single cell methods to define the cellular tropisms of viruses and their genetic programs in the infected cell. This chapter contains descriptions of these methods integrated with illustrations of applications to studies of viral pathogenesis.

2 General Considerations

In situ hybridization can be combined with immunocytochemistry to detect genes and their products in the same cell, or to type cells in which viral nucleic acids

---

[1] Department de Virologie, Institut Pasteur, 28, rue du Dr. Roux, F-75724 Paris Cedex 15

[2] Department of Microbiology, University of Minnesota, Medical School, Minneapolis, Minnesota 55455, USA

* Work in the authors' laboratories is supported by grants from the USPHS, CNRS, INSERM, American Cancer Society, National Multiple Sclerosis Society, Association pour la Recherche sur la Sclérose en Plaques and Institut Pasteur Fondation.

Current Topics in Microbiology and Immunology, Vol. 143
© Springer-Verlag Berlin · Heidelberg 1989

are detected (BRAHIC et al. 1984). In the first case, antibodies react with viral polypeptides, in the second with cellular antigens. Immunocytochemistry is usually performed first with substrates such as diaminobenzidine or pyrecatechol-phenylene-diamine dihydrochloride (PPD) which form polymers that withstand the subsequent treatments of in situ hybridization. The antigens are marked by the brown or black-brown deposits respectively of diaminobenzidine or PPD, and the viral genes by an increased number of autoradiographic silver grains over background. In some instances the order of the assay may be reversed. This improves levels of hybridization and decreases background, probably because the diaminobenzidine or PPD deposits impede access of probes and bind them nonspecifically. In the few examples where this has been successful, the antigens were abundantly represented in the cell and exceptionally stable to aldehyde fixation and in situ hybridization procedures. Riboprobes have also proved to be particularly useful because of the high specific activity of the probe and the low backgrounds achieved with post-hybridization RNase treatment.

Clearly, the methodological problem is to preserve antigenic reactivity under conditions which are compatible with in situ hybridization. The method of fixation, for example, must not only represent the best compromise between diffusion of probe and retention of cellular morphology, it must also preserve the reactivity of relevant epitopes and target sequences. Similarly, the immunocytochemical procedures represent a compromise between detection of antigen, nonspecific binding of probe, and the loss of specific signal due to decreased diffusion and contaminating nucleases. Because changing one parameter will affect all the others, the best conditions must be determined empirically. In the ensuing descriptions of methods and applications, we provide guidelines and conditions from experience which are good starting points.

3 Determining Virus Host Range

Pathological alterations in the white matter, including primary demyelination resembling the lesions of multiple sclerosis, accompany the persistent infections of the central nervous system (CNS) caused by a murine picornavirus, Theiler's murine encephalomyelitis virus (TMEV) and by two lentiviruses, visna virus and human immunodeficiency virus (HIV). The mechanisms of tissue injury fall into two classes in which the oligodendrocyte, the cell in the CNS whose plasma membrane provides the myelin sheath, is either directly involved or not. In the first case, infection of the oligodendrocyte causes altered myelin metabolism or destruction of the cell by the virus or by the immune response mounted against infection (HAASE et al. 1984b). In the second case, inflammatory cells indirectly cause the changes in the white matter through the secretion of proteases and other cytotoxic products. Thus, to understand the mechanism of demyelination, the first question which must be addressed is which type of cell has been infected by the virus. In the sections that follow, we describe the use of the simultaneous detection assay to define viral host range in the CNS.

3.1 Theiler's Murine Encephalomyelitis Virus

Mice inoculated with TMEV are anesthetized and perfused through the left ventricule after an incision is made with scissors in the right atrium. Perfusion is carried out with 20 ml of phosphate-buffered saline (PBS) followed by 20 ml of fixative. The following fixatives have been found suitable for the simultaneous assay: (a) 2%–4% formaldehyde in PBS pH 7.4; (b) 0.5% formaldehyde, 0.5% glutaraldehyde, 0.1 M phosphate buffer pH 6.0, 1.6% glucose, 0.02% $CaCl_2$, 1% dimethylsulfoxide (PFG; MORRIS and BARBER 1983); (c) 2% formaldehyde, 0.075 M lysine, 0.01 M $NalO_4$, 0.037 M phosphate buffer pH 7.5 (PLP; McLEAN and NAKANE 1974); (d) 4% formalin solution, 69% ethanol, 5% glacial acetic acid, 22% H_2O (FEA; CAMMER et al. 1985). When using FEA, PBS is replaced by 0.01 M TRIS HCl pH 7.5, 0.1 M NaCl to avoid intravascular precipitation of phosphate salts.

Following perfusion, the spinal cords are dissected and immersed in fixative at 4 °C for 30 min (the refixation time is increased to 2.5 h in the case of FEA). Fixation by aldehydes is terminated by quenching in cold 0.15 M triethanolamine (pH 7.5) for 30 min followed by two 5-min washes in cold PBS. Tissues are dehydrated and embedded in paraffin [immersion for 1 h each in 75% ethanol, 80% ethanol, 95% ethanol, absolute ethanol twice, xylene twice, and twice in melted paraffin (Paraplast or an equivalent) at 58 °C].

Ten-micrometer-thick sections are cut, floated on warm water (45 °C) and picked up on slides treated with Denhardt's medium (BRAHIC and HAASE 1978), coated with Elmer's white glue (Elmer's Glue-All, Borden, USA) to ensure good adherence of the tissue to the slide. To coat slides with Elmer's glue, deposit one drop of on the slide, smear it with a gloved finger, dip the slide in distilled water obtain an even coat of glue, and let the slide dry for 10–15 min in a vertical position. After drying, the slide should be opalescent. Slides with paraffin sections are dried overnight at 37 °C and can be stored at +4 °C for several months. Prior to immunocytochemistry and in situ hybridization paraffin is removed by dipping three times in xylene and twice in 100% ethanol (5 min each time).

The types of cells infected by TMEV were recognized with antibodies to an astrocyte-specific antigen, glial fibrillary acidic protein (GFAP), on oligodendrocyte-specific antigen, carbonic anhydrase II (CA II) (GHANDOUR et al. 1980; KUMPULAINEN et al. 1983; STERNBERGER 1984), and a surface antigen of microglia and monocytes recognized by monoclonal antibody F4/80 (HUME and GORDON 1983). Two percent formaldehyde or PFG is suitable for the detection of GFAP and the epitope recognized by monoclonal F4/80. Detection of CA II is possible only if tissues are fixed with FEA. For intracellular antigens (GFAP and CA II) diffusion of antibodies is enhanced by dipping tissue sections for 4 min at room temperature in 0.1% Triton X 100 (ROHM and HAAS, Pa.) in PBS followed by two 5-min washes in PBS. Slides are reacted at room temperature with blocking antibody (30 min), PBS (twice 5 min), primary antibody at the appropriate dilution (30 min), secondary antibody coupled to horse-radish peroxidase at a 1:50 dilution (30 min), PBS (5 min), 0.01% H_2O_2, 0,5 mg/ml diaminobenzidine in 0.05 M TRIS HCl (pH 7.5) (5 min), and distilled water (twice 5 min).

Some sera, particularly hyperimmune rabbit antisera, contain high levels of ribonuclease activity which will decrease subsequent detection of viral RNA by in situ

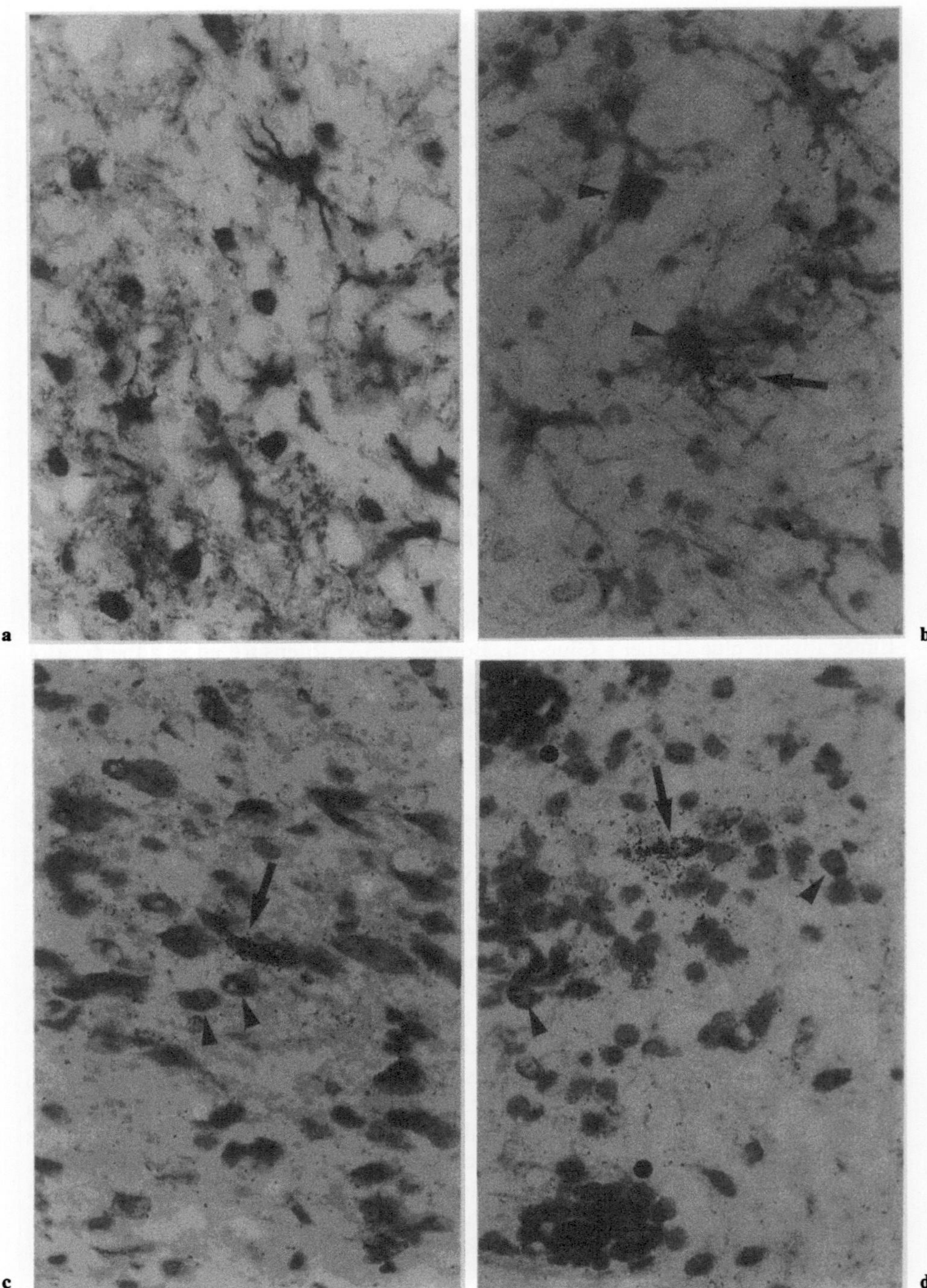

(see also Appendix for color illustration)

hybridization. In this case, sera are diluted in PBS with 2% calf serum and 5 mg/ml heparin, or 0.4% diethylpyrocarbonate, RNase inhibitors which do not interfere with immunocytochemistry (SHIVERS et al., 1986; AUBERT et al. 1987).

When FEA is used as a fixative, RNA secondary is so altered that hybrid formation is impaired. Hybrids will form if the RNA is first denatured in 200 mM glyoxal, 50% dimethylsulfoxide, 10 mM phosphate buffer pH 7.0, 50 °C for 1 h. Tissue sections are then washed three times in distilled water at 4 °C (5 min each time).

Following immunocytochemistry and denaturation of RNA, the sections are pretreated for in situ hybridization in 0.2 N HCl (20 min), 2XSSC at 70 °C (30 min), and proteinase K (1 µg/ml, 37 °C, 15 min). After acetylation to block nonspecific binding of probe to diaminobenzidine, (add 0.25% acetic anhydride to 0.1 M triethanolamine, mix vigorously, and immerse slides for 10 min at room temperature), the sections are washed twice in distilled water and dehydrated by immersing twice in 70% ethanol and once in 90% ethanol. The sections are hybridized with ^{3}H- or ^{35}S-labeled TMEV-specific probe (2 to 3×10^8 dpm/µg), washed, and developed after radioautographic exposures of 7–14 days. Conditions for hybridization and washing are as described (BRAHIC and HAASE 1978; HAASE et al. 1984a).

In Fig. 1, TMEV RNA was detected in cells unequivocally identified as oligodendrocytes. The combined immunocytochemistry–in situ hybridization method showed that 25%–40% of cells containing viral RNA were oligodendrocytes, 10% were astrocytes, and 10% were microglial cells-monocytes. The finding that the oligodendrocyte is the main host cell for TMEV infection speaks for a mechanism of demyelination in which viral infection itself, or the immune response directed towards infected cells, destroys the myelin-forming cell of the CNS.

3.2 Visna

A similar mechanism may operate in the slow infection of sheep caused by visna virus, the prototypic lentivirus. Both the primary and the secondary demyelination which accompany infection can be attributed in part to replication of virus in the oligodendrocyte, again unambiguously identified by the simultaneous detection assay (STOWRING et al. 1985).

The identification of visna virus-infected cells illustrates the case of antigens labile to aldehyde fixatives and to the paraffin embedding procedures. In this case, tissues are frozen between blocks of dry ice, or in isopentane cooled to liquid nitrogen temperature, and stored at −50 °C. After sectioning with a cryostat at −20 °C, the frozen sections are fixed in cold ethanol (4 °C) for 20 min, washed in cold PBS for 20 min, and sequentially incubated at room temperature with 1:200 dilution (in PBS) of normal serum (20 min), 1:100 dilution of a guinea pig polyclonal serum raised against purified sheep oligodendrocyte (20 min), biotinylated

◀ **Fig. 1a–d.** Determining TMEV host range in CNS with combined immunocytochemistry-in situ hybridization. **a** Double immunostaining with anti-GFAP (alkaline phosphatase, *blue*) and anti-CAII (horse radish peroxidase, *brown*) antisera. **b** Combined immunoperoxidase in situ hybridization. The serum used was anti-GFAP. The *arrowheads* point to GFAP positive astrocytes and the *arrow* to an infected cell with TMEV RNA **c**. **d** Combined immunoperoxidase in situ hybridization. The serum used was anti-CAII. *Arrowheads* point to CAII positive oligodendrocytes. The arrow points to infected oligodendrocytes. *Dots* are adjacent to inflammatory infiltrates characteristic of demyelinating lesions. (From AUBERT et al. 1987, with permission)

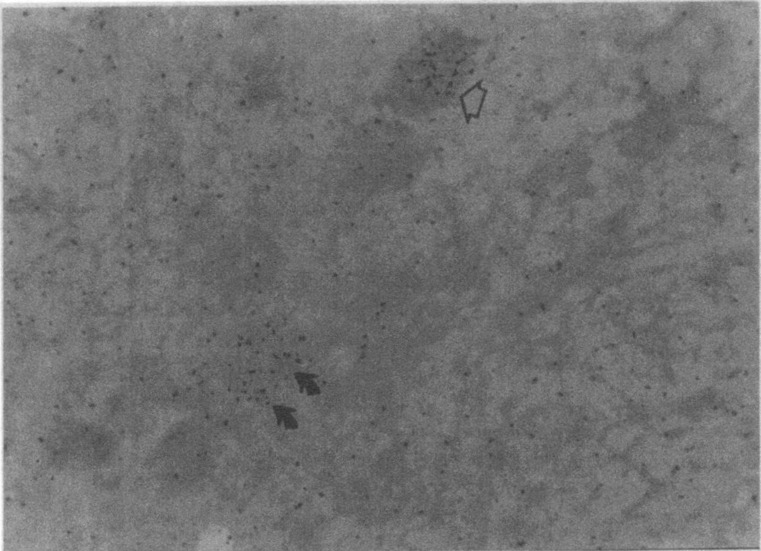

Fig. 2. Visna virus RNA in oligodendrocytes by immunocytochemistry and in situ hybridization. Some of the oligodendrocytes (*open arrow*) marked by the brown diaminobenzidine reaction product contain the virus RNA (*silver grains*), while others do not (*filled arrows*). (From STOWRING et al. 1985, with permission)

(see also Appendix for color illustration)

anti-guinea pig IgG (30 min), 0.3% H_2O_2 in methanol for 30 min (to block endogenous peroxidase in cells such as macrophages or monocytes), avidin biotinylated horseradish peroxidase complex (ABC; 30 min) and diaminobenzidine (0.5 mg/ml) in 0.03% H_2O_2 (5 min). The sections are refixed in ethanol acetic acid (3:1 vol/vol) followed by rinsing in ethanol, treated for in situ hybridization, and after acetylation hybridized to a ^{3}H-radiolabeled visna-specific probe. The sections are coated with NTB-2 emulsion, developed after 3 days, and stained with 0.5% $CuSO_4$ 1% methyl green.

Figure 2 shows visna virus RNA in oligodendrocytes after preparation by the above method. The finding that viral genes are expressed in oligodendrocytes suggests again that demyelination results from direct injury to the cell or from indirect immune-mediated destruction.

3.3 Human Immunodeficiency Virus

The third example of the use of the combined method to analyze host range is drawn from ongoing investigations of the AIDS dementia complex (ADC). More than half of the individuals with AIDS have neurological abnormality, most often associated with pallor of white matter in subcortical areas infiltrated by monocytes, macrophages and giant cells (PRICE et al. 1988). These cells and the resident macrophages in the CNS, microglia, harbor HIV RNA and

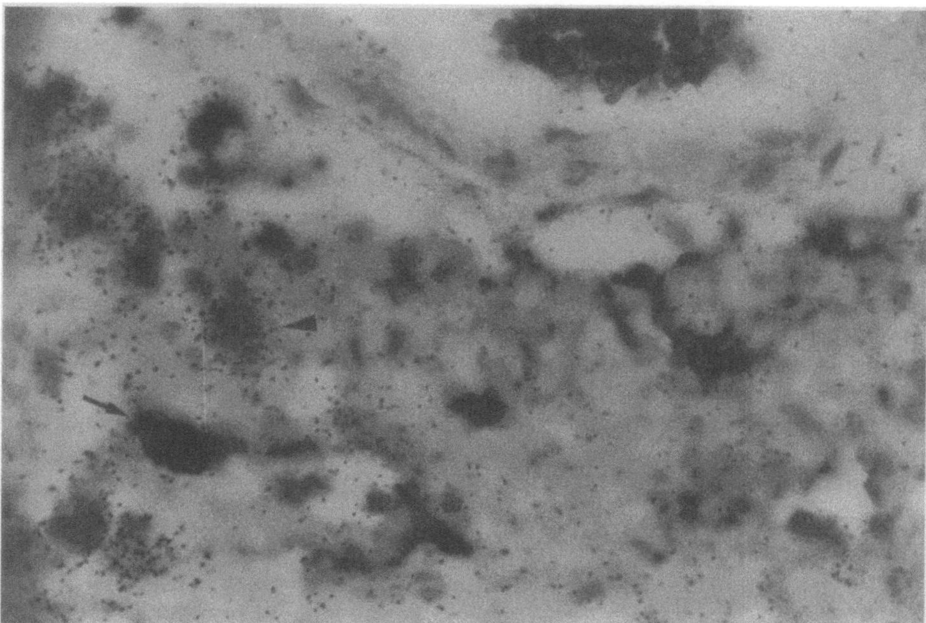

Fig. 3. HIV RNA in cells in the nervous system. Formalin-fixed and paraffin-embedded sections from an individual with AIDS dementia complex were hybridized to a HIV-specific DNA probe radiolabeled by nick translation with ^{35}S precursors. Following hybridization the sections were reacted with antibodies to the astrocyte-specific protein GFAP, biotinylated antispecies antibody, ABC peroxidase, PPD and H_2O_2. Cells of the monocyte lineage (*arrowhead*) in the inflammatory infiltrate bordering the blood vessel at the top of the figure contain HIV RNA whereas the astrocytes (*arrow*) do not. Sample viewed in transillumination and incident polarized light to reveal grains against darker staining background

(see also Appendix for color illustration)

viral antigens (VAZEUX et al. 1987), but it has not been established as yet whether neurons and glial cells harbor virus in a covert state. To evaluate infection of astrocytes, antibodies to GFAP and HIV-specific cDNA probes were used in the simultaneous dectection assay (M. ZUPANIC, R. W. PRICE and A. T. HAASE, unpublished material, 1988). In these studies, probe bound non-specifically and hybridization to cells in the monocyte-microphage lineage were decreased despite acetylation and dilution of antisera to GFAP to the point where staining of astrocytes was barely detectable. These difficulties were overcome by hybrizing first and then reacting the sections with anti-GFAP. GFAP is evidently stable to formalin fixation and in situ hybridization, since astrocytes stained well after in situ hybridization using only a two-fold higher concentration of antibodies than that used for immunocytochemistry performed alone (Fig. 3). Alternatively, with a cRNA probe, immunocytochemistry can be performed first, followed by in situ hybridization. In this case, washing with ribonucleases after hybridization removes nonspecifically bound probe (R. VAZEUX, 1988), personal communication). While these studies cannot as yet exclude latent infection of astrocytes, HIV

RNA was not detected in this cell type, in contrast to the monocytes, microglia, and macrophages, where infection may indirectly lead to cell injury through release of cytotoxins and proteases (PRICE et al. 1988).

4 Viral Genes and Gene Products in the Same Cell: Restricted Gene Expression and Viral Persistence

The viruses responsible for persistent, slow infections of man and animals have evolved mechanisms both to elude host defenses and to moderate destruction of host cells. We have proposed that restriction of viral gene expression could explain both survival of the infected cell and escape from detection by immune surveillance (HAASE et al. 1984b). This theory predicts that infected cells will contain greatly reduced concentrations of viral RNA and antigens when compared with productively infected cells. We now cite two examples where this prediction has been tested with the simultaneous detection assay.

4.1 Theiler's Murine Encephalomyelitis Virus

TMEV persists in the CNS of infected mice despite a vigorous humoral immune response and causes primary demyelination. To examine the hypothesis that TMEV escapes immune elimination through restricted gene expression, sections from chronically infected animals were reacted with antiserum to viral capsid proteins and then hybridized in situ with a TMEV-specific probe. In accord with the hypothesis, the majority of infected cells contained only 100–500 copies of viral

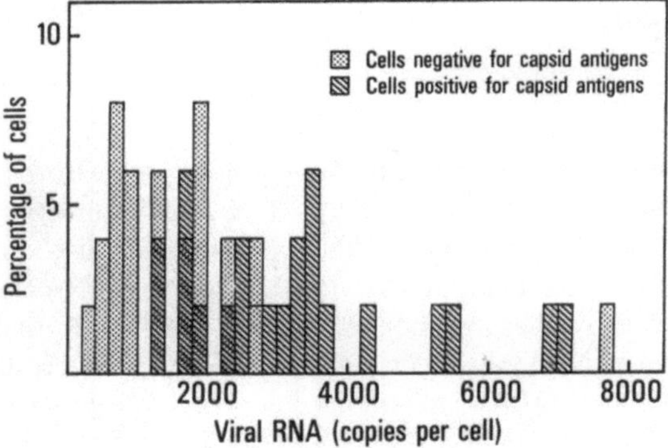

Fig. 4. Distribution of viral RNA and capsid antigens in BHK cells, 6 h after infection with TMEV. Cells were harvested with trypsin, deposited on microscope slides and processed for simultaneous detection of TMEV capsid antigens and RNA with the combined immuno-cytochemistry–in situ hybridization assay. The number of autoradiographic grains per cell was converted to viral RNA copy number with a calibration curve (BRAHIC et al. 1984; CASH et al. 1985). Cells were classified as negative (*shaded bars*) or positive (*open bars*) for viral capsid antigens according to immunoperoxidase staining. The figure is a superimposition of the histrograms obtained for antigen-negative and antigen-positive cells. (From CASH et al. 1985, with permission.)

RNA per cell. Capsid antigens were not detectable because the amount of viral RNA was below a threshold of about 500 copies per cell, (CASH et al. 1985) which has been determined to be the minimal RNA copy number required for synthesis of detectable antigens (Fig. 4). There is, however, a spectrum of states of gene expression, and in cells with higher levels of viral RNA, capsid antigens are also demonstrable.

4.2 Hepatitis B Virus

Infection by hepatitis B virus (HBV) is associated with a broad spectrum of acute and chronic liver diseases and sometimes with hepatocellular carcinoma. The mechanism of persistence in chronic hepatitis was investigated with the simultaneous detection assay applied to formalin-fixed liver sections using antisera to the core antigen of HBV, the presumptive target of the cellular immune response. This work used a modified technique which partially reverses the extensive formaldehyde cross-links and restores the sensitivity of in situ hybridization to allow detection of about one 15-kb target (BLUM et al. 1984).

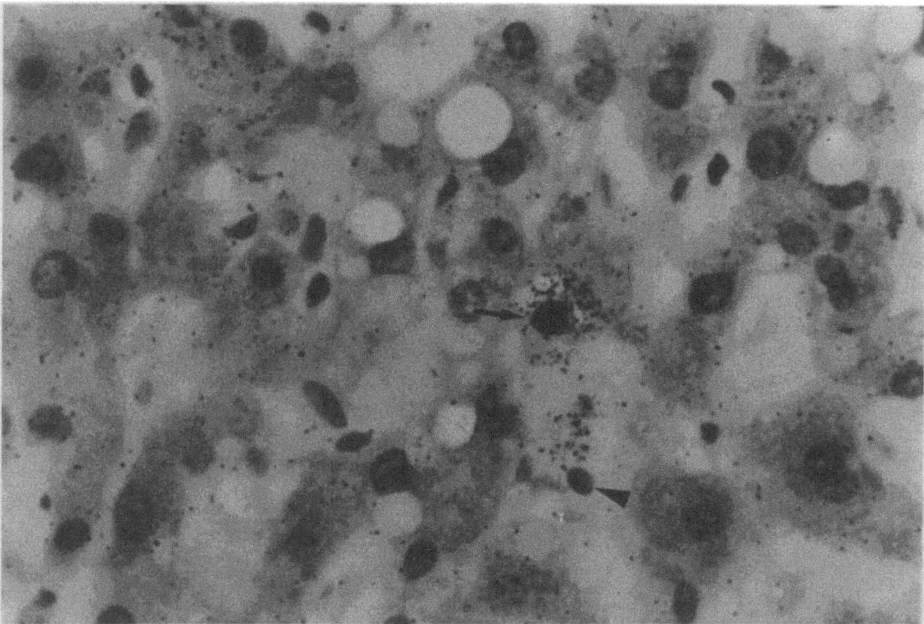

Fig. 5. Simultaneous detection of HBV core antigen and DNA in persistent infections. Formalin-fixed and paraffin-embedded sections of virus from an individual with chronic active hepatitis virus infection were reacted with antibody to HBV core antigen followed by immunocytochemical development using PPD as a chromogen. After the pretreatments described in the text to enhance hybridization, the sections were hybridized to a ^{3}H-labeled HBV-specific probe. The *arrow* points to a cell with HBV core antigens in the nucleus and viral DNA in the cytoplasm. An adjacent cell (*arrowhead*) with much less DNA is not stained for core antigen. (From BLUM et al. 1984, with permission.)

(see also Appendix for color illustration)

Paraffin sections, 5 µm thick, are cut from blocks of liver tissue, picked up on pretreated slides coated with glue, and dried overnight at 37 °C. The sections are deparaffinized by heating for 3 h at 60 °C, followed by three 5-min washes in xylene, and two 5-min washes in 100% ethanol. After rehydration (5 min each in 90% and 70% ethanol, followed by dipping twice in PBS), the sections are incubated with antibody to core antigen, a secondary antibody labeled with peroxidase, and developed with H_2O_2–PPD. Aldehyde cross-links and methylation of amino groups are reversed by treating slides with 0.2 M HCl for 30 min and then with 0.15 M triethanolamine pH 7.4 for 15 min. To enhance diffusion of probe, the sections are incubated at 70 °C for 30 min in 2XSSC, permeabilized with 0.005% digitonin (in 125 mM sucrose, 60 mM KCl, 3 mM Hepes pH 7.4), and digested for 15 min at 37 °C with 5 µg/ml proteinase K in 20 mM TRIS HCl pH 7.4, 2 mM $CaCl_2$. The sections are then hybridized in situ with a radio-labeled HBV-specific probe, washed, coated with NTB-2 emulsion, and developed and stained as described in the preceding sections. Most hepatocytes in which viral DNA is detected (Fig. 5) do not contain core antigen, as would be expected if viral genomes persist in cells in which there is insufficient expression of the antigenic target for elimination by cellular immunity.

5 Double-Label In Situ Hybridization and Color Microradioautography

This chapter concludes with a brief account of a double-label in situ hybridization technique with promises to be useful although there are as yet no specific examples of its application (HAASE et al. 1985). The method exploits the high energy of ^{35}S to penetrate a thin film of clear plastic separating two layers of nuclear track emulsion. When two probes for two genes are hybridized in situ, one labeled with ^{3}H and the other labeled with ^{35}S, only the probe labeled ^{35}S will be recorded in the second layer. To distinguish grains more easily in the two layers grains in the first layer (^{3}H and ^{35}S), are converted to a magenta color, and those in the second (^{35}S only) to a cyan color by a color microradio-autographic method.

As an illustration of the method, cells infected with two different viruses, visna and measles, were mixed and deposited by cytocentrifugation on treated glass slides. After hybridizing to a mixture of a ^{3}H-labeled probe for measles virus RNA and ^{35}S-labeled probe for visna virus RNA, the slides are washed, coated with NTB-2, developed, and treated for 3 min in 0.37% formalin in 0.5% Na_2CO_3 to stabilize emulsion. To convert the grains to a magenta color, slides are washed for 2 min in tap water, bleached for 1 min in 10% $K_3Fe(CN)_6$ 5% KBr, washed again in tap water, and developed for 1 min in freshly prepared dye coupler. The dye coupler is made by dissolving 100 mg Eastman Kodak M-38 in 5 ml ethanol 0.2 N NaOH; and adding this to 45 ml of a solution containing 100 mg of Eastman Kodak color developer CD2 100 mg Na_2SO_3, 50 mg KBr, 1 g Na_2CO_3. The slides are washed, bleached and washed again as above, fixed for 5 min in a solution containing 24 g of $Na_2S_2O_3$ and 15 g of Na_2SO_3 per 100 ml of water. After washing in the water, the slides are then stained with methyl green and coated with a thin film of krylon formed as the solvent evaporates from 0.4 ml krylon dropped

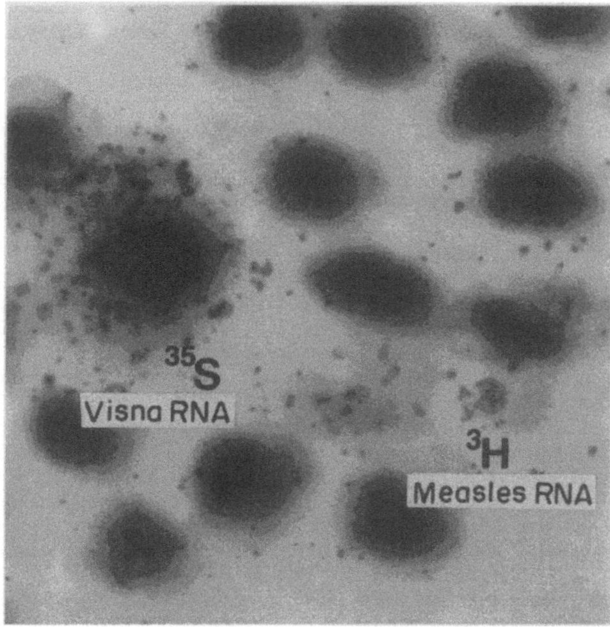

(see also Appendix
for color illustration)

Fig. 6. Double-label in situ hybridization and color microradioautography. Mixed population of cells infected with visna virus or measles virus were hybridized in situ to probes labeled respectively with ^{35}S or ^{3}H. Grains in a first layer of emulsion have been converted to a magenta color, those in the second to a cyan color. Since only ^{35}S penetrates the second layer, the cell with visna virus RNA is identified by the superimposed magenta and cyan colored grains. (From HAASE et al. 1985, reproduced with permission.)

on to water. The slides are dried vertically and coated with 0.5% gelatin 0.55% CrK $(SO_4)_2$, a second layer of emulsion, and exposed for a second time. In the subsequent color development, cyan coupler C16 is substituted for M-38. Figure 6 shows the superimposed magenta and cyan grains over cells infected with visna virus, which bound the ^{35}S probe. Cells with measles virus RNA bound the ^{3}H probe and have only magenta grains.

We expect that this method and the immunocytochemical hybridization assays will find increasing use in viral pathogenesis, developmental biology, neurosciences and other disciplines where defining levels of gene expression in vivo affords deeper insight into mechanism and function.

Acknowledgements. We would like to acknowledge the collaboration of C. Aubert, M. Chamorro, L. Stowring, H. Blum and E. Cash, and thank M. Gau and D. Clark for preparing the manuscript.

References

Aubert A, Chamorro M, Brahic M (1987) Identification of Theiler's virus infected cells in the central nervous system of the mouse during demyelinating disease. Microbial Pathogenesis 3: 319–326

Blum HE, Haase AT, Vyas GN (1984) Molecular pathogenesis of hepatitis B virus infection: simultaneous detection of viral DNA and antigens in paraffin-embedded liver sections. Lancet ii: 771–775

Brahic M, Haase AT (1978) Detection of viral sequences of low reiteration frequency by in situ hybridization. Proc Natl Acad USA 75: 6125–6129

Brahic M, Haase AT, Cash E (1984) Simultaneous in situ detection of viral RNA and antigens. Proc Natl Acad Sci USA 81: 5445–5448

Cammer W, Sacchi R, Sapirstein V (1985) Immunocytochemical localization of carbonic anhydrase in the spinal cords of normal and mutant (shiverer) adult mice with comparisons among fixation methods. J Histochem Cytochem 33: 45–54

Cash E, Chamorro M, Brahic M (1985) Theiler's virus RNA and protein synthesis in the central nervous system of demyelinating mice. Virology 144: 290–294

Ghandour MS, Langley OK, Vincendon G, Gombos G, Filippi D, Limozin N, Dalmasso C, Laurent G (1980) Immunochemical and immunohistochemical study of carbonic anhydrase II in adult rat cerebellum: a marker for oligodendrocytes. Neuroscience 5: 559–571

Haase AT, Brahic M, Stowring L, Blum H (1984a) Detection of viral nucleic acids by in situ hybridization. Methods Virol 7: 189–226

Haase AT, Pagano J, Waksman B, Nathanson N (1984b) Detection of viral genes and their products in chronic neurological diseases. Ann Neurol 15: 119–121

Haase, AT, Walker D, Stowring L, Ventura P, Geballe A, Blum H, Brahic M, Goldberg R, O'Brien K (1985) Detection of two viral genomes in single cells by double-label hybridization in situ and color microradioautography. Science 227: 189–192

Hume DA, Gordon S (1983) Mononuclear phagocyte system of the mouse defined by immuno-histochemical localization of antigen F4/80. J Exp Med 157: 1704–1709

Kumpulainen T, Dahl D, Korhonen LK, Nyström SAM (1983) Immunolabeling of carbonic anhydrase isoenzyme C and glial fibrillary acidic protein in paraffin-embedded tissue sections of human brain and retina. J Histochem Cytochem 22: 1077–1083

McLean IW, Nakane PK (1974) Periodate-lysine-paraformaldehyde fixative. A new fixative for immunoelectron microscopy. J Histochem Cytochem 22: 1077–1083

Morris RG, Barber PC (1983) Fixation of Thy-1 in nervous tissue for immunohistochemistry. J Histochem Cytochem 31: 263–274

Price RW, Brew B, Sidtis J, Rosenblum M, Scheck AC, Cleary P (1988) The brain in AIDS: central nervous system HIV-1 infection and AIDS dementia complex. Science 239: 586–592

Shivers BD, Harlan RE, Pfaff DW, Schachter BS (1986) Combination of immunocytochemistry and in situ hybridization in the same tissue section of rat pituitary. J Histochem Cytochem 34: 39–43

Sternberger NH (1984) Patterns of oligodendrocytes function seen by immunocytochemistry. In: Norton WT (ed) Oligodendroglia. Plenum, New York, pp 125–173

Stowring L, Haase AT, Petursson G, Georgsson G, Palsson P, Lutley R, Roos R, Szuchet S (1985) Detection of visna virus antigens and RNA in glial cells in foci of demyelination. Virology 141: 311–318

Vazeux R, Brousse N, Jarry A, Henin D, Marche C, Vedrenne C, Mikol J, Wolff M, Michon C, Rozenbaum W, Bureau JF, Montagnier L, Brahic M (1987) AIDS subacute encephalitis: identification of HIV infected cells. Am J Pathol 126: 403–410

Colorimetric In-Situ Hybridization in Clinical Virology: Development of Automated Technology

E. R. Unger[1] and D. J. Brigati[2]

1 Introduction

The goals of clinical virology, to generate accurate, sensitive and specific diagnoses as rapidly as possible, have not changed, but the methods to achieve these goals have undergone a revolution. The same techniques of molecular biology which proved so valuable in elucidating basic biologic properties of the viruses have been found to be applicable in a diagnostic setting. Through the adoption of monoclonal antibody and nucleic acid hybridization techniques, the clinical virology lab has been the indicator of things to come in microbiology in general. Because the culture of viruses is labor-intensive, expensive and often slow, the laboratory has been eager to adopt alternative methods to complement and streamline the "gold standard" of viral culture. The growing recognition of the clinical significance of noncultivatable viruses such as rota virus and human papilloma virus has also given impetus to the application of monoclonal antibodies and nucleic acid hybridization to these problems.

The development of stable nonradioactive affinity-labeled nucleic acid probes (LANGER et al. 1981) which could be detected colorimetrically with sensitivity approaching that of radioactive-based methods (LEARY et al. 1983) gave great impetus to the introduction of hybridization-based technology to the clinical laboratory. Optimism about the potential clinical applications of this technology

[1] Department of Pathology and Laboratory Medicine, Emory University, 1364 Clifton Road, Atlanta, GA 30322, USA
[2] Department of Pathology, The Milton S. Hershey Medical Center, The Pennsylvania State University, Hershey, PA 17033, USA

Current Topics in Microbiology and Immunology, Vol. 143
© Springer-Verlag Berlin · Heidelberg 1989

was high, and research news reviews as early as September 1983 predicted widespread diagnostic applications (LEWIN 1983). Detection of adenovirus and cytomegalovirus by colorimetric in situ hybridization in formalin-fixed paraffin-embedded tissue sections (BRIGATI et al. 1983) provided the first example of the clinical application of this new technology. The September 1985 issue of *Clinics in Laboratory Medicine* was devoted to reviews of nucleic acid probes and monoclonal antibodies as diagnostic tools. Arguments as to the usefulness of hybridization technology in the diagnostic setting have been made in many different reviews (e.g., GRODY et al. 1987; HIGHFIELD and DOUGAN 1985; LANDRY and FONG 1985, PALVA 1986; SKLAR 1985; UNGER et al. 1987) and need not be repeated here; the potential diagnostic usefulness of this technology has been widely accepted. Yet despite this acceptance, diagnostic nucleic acid hybridization remains a potential rather than a reality for most clinical laboratories.

What factors have prevented the widespread adoption of this technology which has been, and is, a source of much optimistic discussion between research scientists, pathologists and clinicians? Certainly the field would benefit from commercial availability of well-characterized affinity-labeled probes and from improved colorimetric detection methods giving higher sensitivity. Yet, as important as advancements in these areas are, they would do nothing to resolve the biggest impediment to widespread adoption of hybridization methods in the clinical laboratory: the technical burden of manual sample preparation and processing. As noted by HIGHFIELD and DOUGAN (1985), "A crucial area for development is in the format of the assay ... The techniques so far described are not suitable for large numbers of samples."

At least some form of automation will be required to allow the cost-effective efficient handling of the multiple samples needed in the clinical setting. PALVA (1986) listed five requirements of a practical hybridization test; minimal pretreatment of samples, speed, maximal sensitivity, nonisotopic detection and *automation*. While the need for automated hybridization methods is clear, a recent literature search indicated a total absence of publications in this field. This review will address some of the requirements for development of an automated hybridization assay and indicate the steps which have been taken toward making this a reality.

2 Automation of Hybridization Assay

2.1 Assay Format: Selection of In Situ Hybridization

Many different hybridization assays have been described and found useful in various diagnostic situations. These include dot blot, sandwich, and Southern/Northern hybridizations, as well as in situ hybridization. Each of these assays differ in the sample preparation, specimen requirements and sensitivity. Our laboratory, based in surgical pathology, has found in situ hybridization to be most suitable to our samples and diagnostic questions.

One of the chief appeals of in situ hybridization is the potential for complementing the diagnostic information available in the morphologic detail with more specific genetic information. The morphologic preservation which is central to the method allows determination of which cells contain the genetic sequences of interest as well as where in the cell these sequences occur. The morphologic context facilitates interpretation of results because the pattern of signal can be more easily distinguished from the background or nonspecific pattern. Nonspecific hybridization has been described to be a significant problem in some circumstances (AMBINDER et al. 1986). In situations where rare highly infected cells are expected in a negative background, extraction-based techniques give negative or equivocal results because of signal dilution. In situ hybridization formats the results in a tissue or cellular context which obviates this problem and results in increased sensitivity and specificity. In situ hybridization utilizes routinely processed formalin-fixed paraffin-embedded tissue sections, frozen tissue sections or cells as the starting material. While this involves some degree of sample preparation, the techniques are already in routine use in standard histology and cytology laboratories.

In some diagnostic settings the sample does not include morphologically intact tissue or cells (e.g., the demonstration of rota virus in fecal samples). In these cases an in situ hybridization assay is neither desirable nor feasible and another format should be chosen. (However, hybridization of an intestinal mucosal biopsy could be substituted for fecal sampling.) Each assay format will impose unique constraints on the design of an automated system. While this review is restricted to a consideration of the in situ hybridization assay, many of the same factors would be involved in the automation of other hybridization methods.

2.2 Optimization of Assay

One of the crucial lessons of automation is that it is not a solution to methodological problems. A complex assay which is time consuming, technically difficult and inherently unstable will not be greatly enhanced by automation. Therefore, one of the most important steps toward automation of in situ hybridization was to streamline and simplify the method without reducing the sensitivity of the result. We have published a method of in situ hybridization (UNGER et al. 1986) which utilizes formalin-fixed paraffin-embedded tissue sections and yields a colorimetric result within 8 h. This reduced the time and complexity of the previously metric method (BRIGATI et al. 1983) without significantly compromising sensitivity.

This streamlined approach involves protease predigestion of the tissue, nick-translated biotin-labeled DNA probes, simultaneous heat denaturation of probe and tissue DNA, and direct detection of specifically hybridized biotinylated probe with an avidin-alkaline phosphatase complex. Other workers have subsequently published very similar methods using other conditions of protease digestion or enzymatic detection (e.g., BURNS et al. 1986; LEWIS et al. 1987; SINGER et al. 1986).

Equally important to method optimization is an understanding of which steps of the assay are most crucial. While each step of the assay must be performed properly to insure success, we have found the steps of protease digestion and denaturation to be most crucial. Both of these steps are empirically determined and greatly influenced by the degree and kind of fixation of the starting material.

When considering the automation of this assay, special attention should be given to these two crucial steps.

2.3 Requirements of Automated System

The ideal automated system for in-situ hybridization should:

1. Not consume any more precious reagents or sample per assay than manual methods.
2. Utilize the same starting material and be as rapid and reproducible as manual methods.
3. Have sufficient flexibility to adapt to the changing technology of the hybridization assay.
4. Perform all steps of the assay irrespective of the complexity, order, time or temperature required.
5. Have minimal maintenance requirements and add stability to the assay.
6. Be capable of running without supervision.
7. Be completely contained in a safety hood and use no known or suspect carcinogens.
8. Be interfaced with state-of-the-art, user-friendly computer technology capable of rapid data collection and reprogramming by laboratory personnel with standard training.
9. Conserve valuable technician time well in excess of the best available manual technology.
10. Be competitively priced commensurate with the technician time saved.

The first requirement of an automated assay, to conserve valuable reagents, is essential for a practical economic method. The solution to this volume constraint will influence the configuration of the assay and thus the volume becomes a central barrier to the development of a successful automated system. Many of these same constraints, including the volume barrier, have been addressed in the automation of immunohistochemistry (BRIGATI et al. 1988) and resulted in the development of capillary gap technology by one of us (DJB) as well as in the commercial production of the Fisher Code-On Stainer Fisher Scientific, Pittsburgh, PA, USA interfaced with an IBM System II personal computer. Lessons learned from the automation of immunochemistry can be effectively applied to the automation of in situ hybridization.

2.4 Capillary Gap Technology

Immunohistochemistry and in situ hybridization share many similarities, and, when broken down to their component steps, both involve the timed application and removal of a series of reagents. The conditions of interaction of reagent and tissue substrate may vary at each step and each reagent may be applied at more than one step. Some reagents are common to all tissues run in the assay, while other reagents are discrete and possibly different for each tissue. Small volumes of precious reagents (of the order of 100 µl) must evenly cover the tissue wherever it is placed on the slide.

Capillary gap technology proved central to the automation of the component steps of immunocytochemistry (BRIGATI et al. 1988). The gap produced by sand-

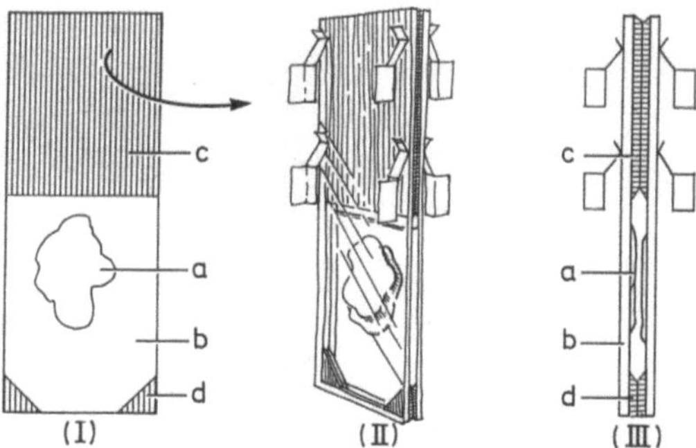

Fig. 1. Capillary action slides as seen in three different views. *I* Front view shows a tissue section (**a**) immobilized in the clear glass space (**b**) below a 75-μm-thick painted rectangle (**c**) and above the two 75-μm-thick painted triangles (**d**) located at each bottom corner. *II* When the painted portions of two capillary action slides are apposed so that their matching painted areas touch, a 150-μm gap is automatically formed. *III* Side view shows that tissue sections (**a**) can be immobilized on both slides that form the capillary gap without impeding filling or emptying. The 150-μm gap is maintained by the combined thickness of the painted areas on the glass slides (**c**) and (**d**)

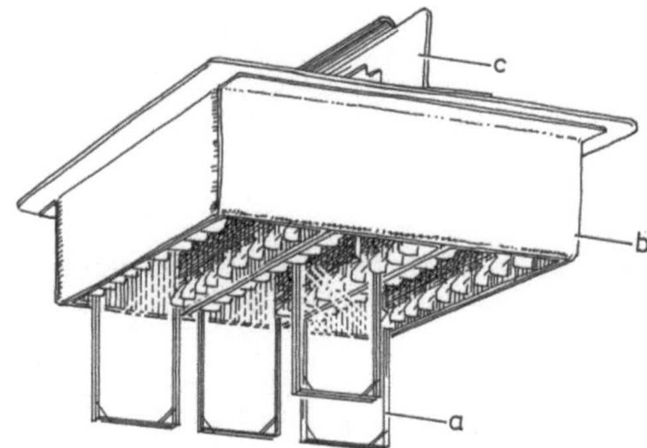

Fig. 2. The capillary action slides (**a**) are immobilized in a special holder (**b**). This holder immobilizes 30 pairs of capillary action slides simultaneously and it is moved by the Fisher Code-On Stainer by its attachment handle (**c**). Each capillary gap can receive a unique set of reagents

wiching two standard microscopic slides and a one and one-half coverslip is approximately 150 μm wide. The coverslip may be replaced by a 75 μm coat of paint on the slides (Fig. 1). When such a sandwich is touched to a liquid surface, the gap fills spontaneously by capillary action. A volume of 150 μl is sufficient to cover the lower 75% of both slides' inner surfaces. Tissue sections 3- to 5-μm thick immobilized anywhere on the lower 75% of the inner surface of one or both slides forming the gap, do not interfere with filling by capillary action. Thus the maximum volume of reagent required to insure an even covering of the sections is 75 μl per slide.

When placed in a special slide holder (Fig. 2), 30 slide pairs (i.e., 60 slides) are conveniently and efficiently handled in minimum space. The vertical arrangement keeps the lower ends of the gaps in one plane and all reagents fill from this bottom plane. For common steps of the assay when all gaps are treated with the same reagent, the reagent can be placed in a bucket or shallow tray.

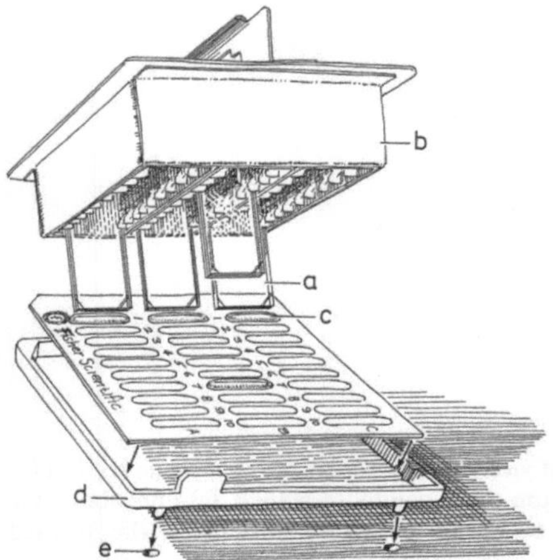

Fig. 3. The mechanism of reagent application by capillary action. The capillary gaps (**a**) are immobilized in their special holder (**b**) and are lowered onto a rubber reagent isolator (**c**). This reagent isolator allows 30 caplet-shaped wells to be filled with microliter quantities of reagents which remain separate from each other because of the hydrophobicity of the isolator surface. The reagent isolator in this illustration has four wells filled with different reagents. The arrangement of these wells matches the pattern of the slides in the holder, so that each capillary gap touches its own well when the holder is lowered onto the reagent isolator. The isolator fits securely into a plastic holder (**d**) which is indexed into the instrument by the position of the recesses (**e**) in the base of the Code-On Stainer. The reagents enter the gaps by capillary action

When the holder enters a bucket, the brim of the holder, which extends beyond the slides, contacts the top of the bucket. The holder-bucket unit can then function as an incubation chamber.

A novel reagent format was designed to allow each capillary gap to receive a minimal volume of discrete reagents (Fig. 3). This reagent isolator has 30 discoid depressions in a planar hydrophobic surface, arranged in a grid pattern exactly matching the spacing of the capillary gaps when maintained in the holder. When 150–200 µl aliquots · of water-based reagents are placed in the recesses, surface tension prevents mixing of adjacent reagents and causes each reagent to stand above the surface as a droplet. The holder with arrayed pairs of slides can be lowered to allow each gap to contact one unique position and thus receive one specified unique reagent. Each gap fills simultaneously with a minimum volume of unique reagent.

Liquid spontaneously fills the gap, but will not spontaneously exit. The slides and confined reagents may be transported to a temperature-controlled chamber during the incubation. Reagent will interact with the tissues until removed from the gap by absorption (Fig. 4). Thus the time of each step may be precisely controlled. It is clear that this capillary gap cycle — the cycle of reagent entry, retention and removal — could form the basis of successful robotic liquid handling. Automation would require a means of transferring the slides in the holder through the capillary gap cycle with the proper reagent sequence and timing.

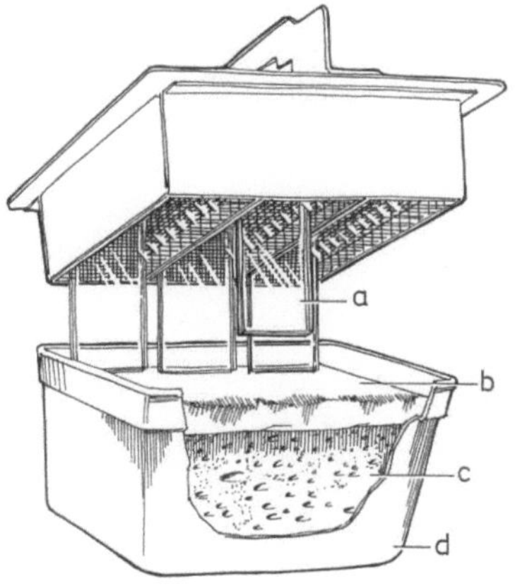

Fig. 4. The mechanism of reagent removal. Removal of the reagents in the capillary gaps (**a**) is achieved by robotically placing the bottom edges of the slides on a blotter pad (**b**). The blotter pad is supported by a sponge material (**c**) in a plastic 400-ml container (**d**). The blotter pad adsorbs the reagent and removes it by reverse capillary action

Fig. 5. The automated in situ DNA hybridization unit. A Code-On Hybridization System consists of an IBM Personal System II computer which runs a tabletop robotic unit (on the *right*) through a central interface box. A printer allows the unit to develop hard copies of the programs and a modem allows laboratories to share software developments, data, and word processing functions

The Fisher Histomatic Model 172 Slide Stainer, a programmable tabletop robot consisting of a mechanical arm capable of moving a slide holder in the x,y,z axes to 19 different reagent stations in a nonlinear fashion, was modified to allow automation of the immunochemistry assay. This relatively inexpensive, mechanically reliable instrument was self-contained in a charcoal filtered hood and had many desirable robotic qualities. Control of the mechanical arm was transferred to an IBM personal computer to facilitate programming of the multiple steps of the assay (Fig. 5). One of the original 19 stations of the Histomatic Slide Stainer was changed to an incubation chamber with computer-adjusted temperature control between room temperature and 60 °C. An interactive program was developed to simplify the

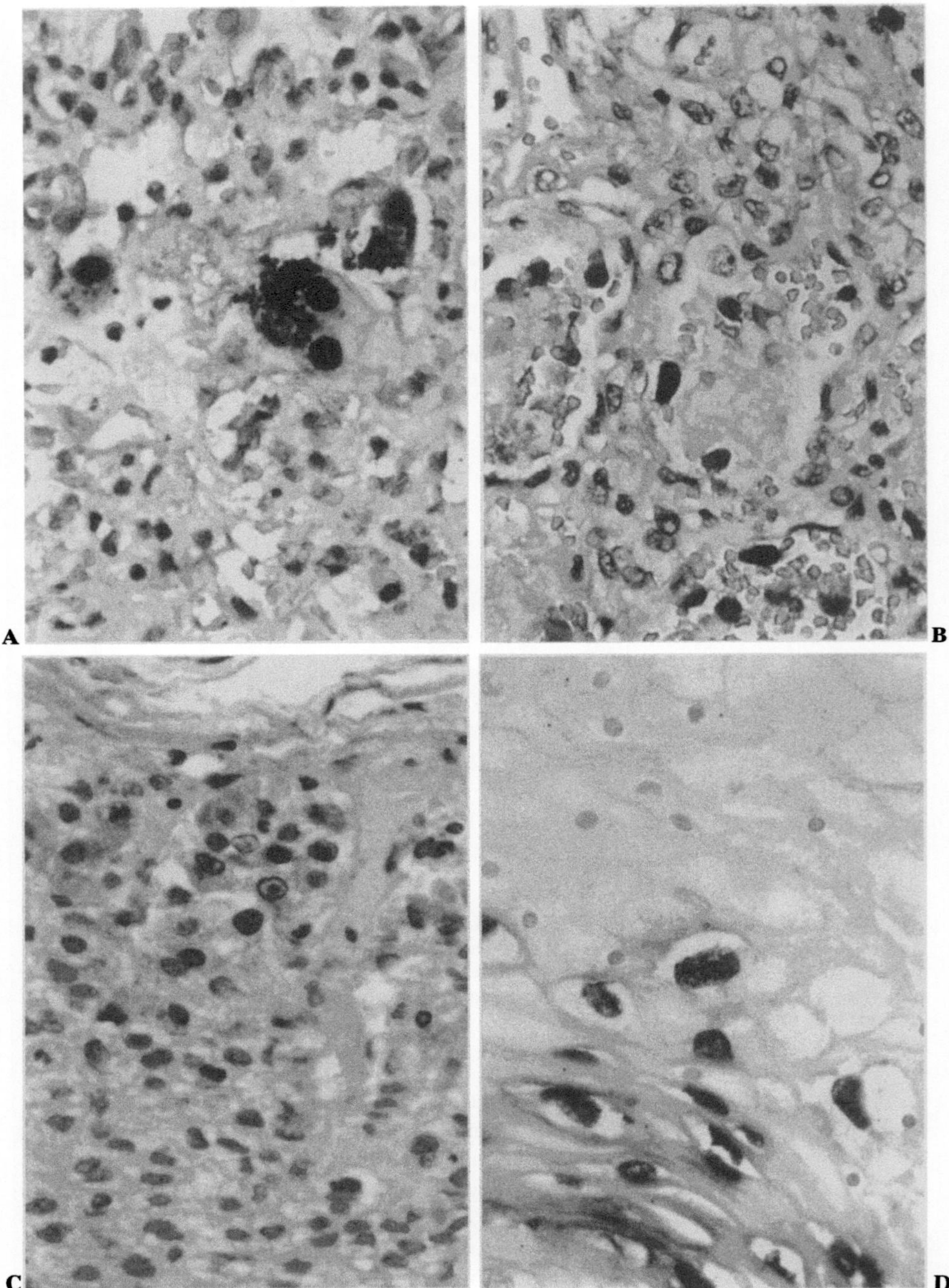

(see also Appendix for color illustration)

calibration of each reagent station position and the development and storage of immunochemical assays. This system is now commercially available as the Fisher Code-On Immunology System and is the first practical automated system for immunocytochemistry (BRIGATI et al. 1988).

2.5 Modifications Required for In Situ Hybridization

Our laboratory felt that the Fisher Code-On Immunology System could be successfully modified to perform in situ hybridization. The automation had the necessary flexibility to perform the number of steps required, as each assay can have any number of events and the time of each event may be varied from 0.00 to 99 min. The biggest differences between immunochemistry and in situ hybridization, from the instrumentation point of view, are the increased viscosity of the hybridization cocktail and the necessity for a denaturation step. The viscosity problem is easily overcome by allowing increased time for the hybridization cocktail to flow in or out of the gap. Our laboratory and the engineers at Fisher are collaborating in development of an additional station with programmable temperature control between room temperature and 110 °C. This oven station should allow for the simultaneous heat denaturation of the tissue and probe DNA in a similar fashion to that achieved in the convection oven with our manual method (UNGER et al. 1986). Our manual assay could be transferred fairly directly to the automated instrument. Times of each step have been found to be similar for both the manual and automated technology. The digestion step is able to be more precisely controlled with time and temperature. Preliminary results are very encouraging and automation of colorimetric in situ hybridization should soon become a reality.

2.6 Applications

The automation of in situ hybridization will greatly facilitate the transfer of hybridization technology from the research laboratory to the diagnostic pathology laboratory. The specific diagnosis of viral infections in the setting of acquired immune deficiency (AIDS), immunosuppressed organ transplant patients and cancer patients is of increasing clinical importance. The most important viruses in these clinical settings are cytomegalovirus, herpes simplex virus and adenovirus. Examples of colorimetric in situ hybridization demonstrating the genes of these viruses are shown in Fig. 6A–C.

◄ **Fig. 6A–D.** Colorimetric in situ hybridization in formalin-fixed paraffin-embedded tissue sections. Method modified from UNGER et al. 1986. **A** Cytomegalovirus in open lung biopsy from bone marrow transplant patient. Biotinylated cloned cytomegalovirus probes detected with avidin-alkaline phosphatase and McGadey reagent. Signal is dark brown/black. Hematoxylin counterstain. **B** Adenovirus in lung from autopsy of child with severe combined immune deficiency following bone marrow transplantation. Biotinylated whole genomic adenovirus 2 DNA detected as in **A**. Hematoxylin counterstain. **C** Herpes simplex virus (HSV) detected in adrenal gland of a congenitally infected infant. Biotinylated cloned HSV probe detected with avidin-alkaline phosphatase and modified McGadey reagent (iodonitrotetrazolium substituted from nitroblue tetrazolium). Signal is red/brown. Hematoxylin counterstain. **D** Human papilloma virus type 6 detected in cervical biopsy of patient with persistent atypia on pap smear and coloposcopic abnormality. Biotinylated cloned HPV 6 probe detected as in **A**. Neutral fast red counterstain

The diagnosis and typing of human papilloma virus (HPV) infection requires hybridization technology. At this time the evidence suggests that patients infected with certain HPV types are at greater risk of malignant disease and therefore warrant more careful clinical monitoring than the general population. The diagnosis and typing of HPV can be achieved with colorimetric in situ hybridization (Fig. 6D) and this area represents another area of application of this technology.

3 Summary

Automation of in situ hybridization is an important first step toward practical implementation of the widely recognized diagnostic potential of nucleic acid hybridization. Our laboratory has concentrated its efforts towards automating colorimetric in situ hybridization on formalin-fixed paraffin-embedded tissue sections. We have capitalized upon the technology developed for the automation of immunohistochemistry (BRIGATI et al. 1988) and are collaborating with Fisher Scientific in modifying the Fisher Code-On Stainer to achieve successful automated in situ hybridization. Preliminary results are encouraging. We feel that the capillary gap technology has the potential to be modified to automate other hybridization assay formats such as dot and sandwich blot hybridizations. While specifically developed for colorimetric hybridization, the instrumentation is self-contained and could be safely adapted to the use of radio-labeled probes if necessary.

References

Ambinder RF, Charache P, Staal S, Wright P, Forman M, Hayward SD, Hayward GS (1986) The vector homology problem in diagnostic nucleic acid hybridization of clinical samples. J Clin Microbiol 24: 16–20

Brigati DJ, Myerson D, Leary JJ, Spalholz B, Travis SZ, Fong CKY, Hsiung GD, Ward DC (1983) Detection of viral genomes in cultured cells and paraffin-embedded tissue sections using biotin-labeled hybridization probes. Virology 126: 32–50

Brigati DJ, Budgeon LR, Unger ER, Koebler D, Cuomo C, Kennedy T, Perdomo JMI (1988) Immunocytochemistry is automated. The development of a robotic workstation based upon the capillary action principle. J Histotechnol 11: 165–183

Burns J, Redfern DRM, Esiri MM, McGee JO'D (1986) Human and viral gene detection in routine paraffin embedded tissue by in situ hybridisation with biotinylated probes: viral localisation in herpes encephalitis. J Clin Pathol 39: 1066–1073

Grody WW, Cheng L, Lewin KJ (1987) In-situ viral DNA hybridization in diagnostic surgical pathology. Hum Pathol 18: 535–543

Highfield PE, Dougan G (1985) DNA probes for microbial diagnosis. Med Lab Sci 42: 352–360

Landry ML, Fong CKY (1985) Nucleic acid hybridization in the diagnosis of viral infections. Clin Lab Med 5: 513–529

Langer PR, Waldrop AA, Ward DC (1981) Enzymatic synthesis of biotin-labelled polynucleotides: novel nucleic acid affinity probes. Proc Natl Acad Sci USA 18: 6633–6637

Leary JJ, Brigati DJ, Ward DC (1983) Rapid and sensitive colorimetric method for visualizing biotin-labeled DNA probes hybridized to DNA or RNA immobilized on nitrocellulose: Bio-blots. Proc Natl Acad Sci USA 80: 4045–4049

Lewin R (1983) Genetic probes become even sharper. Science 229: 1167

Lewis FA, Griffiths S, Dunnicliff R, Wells M, Dudding N, Bird CC (1987) Sensitive in situ hybridisation technique using biotin-streptavidin-polyalkaline phosphatase complex. J Clin Pathol 40: 163–166

Palva A (1986) Microbial diagnostics by nucleic acid hybridization. Ann Clin Res 18: 327–336

Singer RH, Lawrence JB, Villnave C (1986) Optimization of in situ hybridization using isotopic and non-isotopic detection methods. Biotechniques 4: 230–250

Sklar J (1985) DNA hybridization in diagnostic pathology. Hum Pathol 16: 654–658

Unger ER, Budgeon LR, Myerson D, Brigati DJ (1986) Viral diagnosis by in-situ hybridization: description of a rapid simplified colorimetric method. Am J Surg Pathol 10: 1–8

Unger ER, Leary JJ, Ward DC, Brigati DJ (1987) Application of nucleic acid hybridization in clinical virology. In: Microbial antigenodiagnosis, vol 1. CRC Press, Boca Raton

Whole Animal Section In Situ Hybridization and Protein Blotting: New Tools in Molecular Analysis of Animal Models for Human Disease

W. I. Lipkin[1], L. P. Villarreal[2], and M. B. A. Oldstone[1]

1 Introduction

Laboratory animals continue to be invaluable in the study of the pathogenesis and management of infectious diseases. Development of nucleic acid and antibody technologies have facilitated analysis of animal infection models at the levels of genes, messages and proteins. With the advent of whole animal sectioning (WAS), it is now possible to study these models at the molecular level in an anatomic context. WAS in situ hybridization and protein blotting is a sensitive and specific approach to analysis of infections with several viruses, including ground squirrel hepatitis virus, lymphocytic choriomeningitis virus, polyoma virus, rabies virus, reovirus and vesicular stomatitis virus. A similar technique, whole organ sectioning, has been useful for analysis of viral infections in larger hosts, e.g., hepatitis B virus of liver and measles virus infection of brain in man and visna virus infection of brain in sheep (Table 1).

Our general method for WAS in situ hybridization and protein blotting is described below and illustrated in Fig. 1 (for detailed protocols the reader is

[1] Department of Immunology, Research Institute of Scripps Clinic, 10666 No. Torrey Pines Road, La Jolla, CA 92037, USA
[2] Department of Molecular Biology and Biochemistry, University of California (Irvine), Irvine, CA 92717, USA
This is Publication Number 5014-IMM from the Department of Immunology, Research Institute of Scripps Clinic, La Jolla, CA 92037. This work was supported in part by United States Public Health Service grants NS-12428, AG-04342, AI-07007, AI-09484 (MBAO) and GM-36605 (LPV). WIL is a recipient of a Clinical Investigator Development Award from the National Institute of Neurological and Communicative Disorders and Stroke (NS-01026).

Table 1. Infections studied in macroscopic sections by in situ hybridization or protein blotting

Ground squirrel hepatitis virus (LIPKIN et al. 1988, this chapter, Fig 7)
Hepatitis B virus (HAASE et al. 1985b)
Lymphocytic choriomeningitis virus (LIPKIN and OLDSTONE 1986; SOUTHERN et al. 1984; OLDSTONE et al. 1986; BLOUNT et al. 1986)
Measles (HAASE et al. 1985a)
Polyoma virus (ROCHFORD et al. 1987; DUBENSKY et al. 1984)
Rabies virus (BLOUNT et al. 1986)
Reovirus (LIPKIN and OLDSTONE 1986)
Vesicular stomatitis virus (DUBENSKY et al. 1984)
Visna virus (BRAHIC et al. 1981; HAASE et al. 1982)

referred to previous publications from our laboratories: DUBENSKY et al. 1984; SOUTHERN et al. 1984; BLOUNT et al. 1986; LIPKIN and OLDSTONE 1986). Succeeding sections show how WAS has been used to map the distribution of viruses and viral variants, chart the course of natural infections, follow the course of infections after experimental intervention, and examine the effects of infection on expression of host genes and gene products.

2 Materials and Methods

Animals are anesthetized for sacrifice through deep axillary incision. They are then shaved and frozen into blocks of 3.5% carboxymethyl cellulose in a bath of dry ice – ethanol (Fig. 1b). Cryomicrotome sections 20–40 μm thick are collected on transparent adhesive tape (# 688 3M) as shown in Fig. 1c–e and either transferred immediately to membranes for protein blotting or allowed to thaw and dry for 10–30 min at room temperature for in situ hybridization.

2.1 In Situ Hybridization

After sections on tape have air-dried, they are fixed in freshly prepared 4% paraformaldehyde in phosphate buffered saline (PBS), pH 7.4, for 30 min. After fixation, sections are washed in PBS and either used immediately for in situ hybridization or stored for as long as several months, tissue side down against clean glass plates, at 4 °C.

Although both cDNA and cRNA probes have been used, in our view cDNA probes are preferable because hybridization and washes are performed at lower temperatures and sections are less likely to detach from tape. cDNA probes are labeled with ^{32}P either by nick translation (Rigby et al. 1977) or random hexanucleotide primer reactions (Feinberg and Vogelstein 1983) to a specific activity of $1–5 \times 10^8$ cpm/μg of DNA.

Sections are prehybridized for 4–12 h in 50% formamide, $5 \times$ SSC, $2.5 \times$ Denhardt's solution (Maniatis et al. 1982), 150 μg/ml boiled, sonicated salmon sperm DNA. They are then hybridized for 24–48 h in freshly prepared prehybridization solution

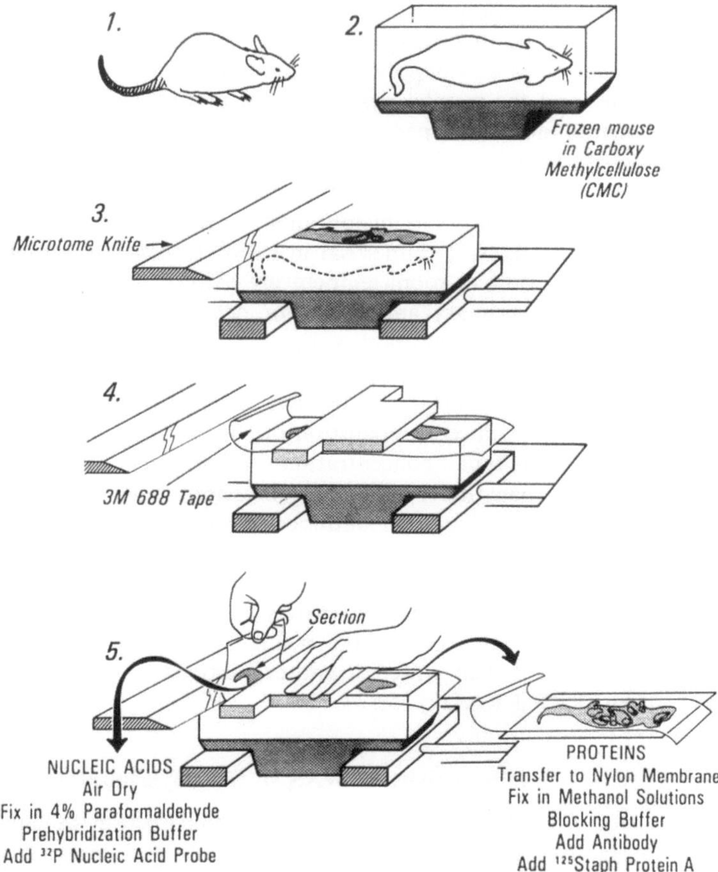

Fig. 1. Method for preparation of animals for cryomicrotomy, whole-animal section in situ hybridization and protein blotting. *1, 2* Animals are sacrificed and frozen in blocks of carboxymethyl cellulose mounting medium *3–5* 30 μm sagittal sections are cut with a cryomicrotome, picked up on tape and processed either for in situ hybridization with probes to viral or host nucleic acids or for protein detection with antibodies to viral or host antigens. (From BLOUNT et al. 1986)

to which probe has been added to a concentration of 1–2×10^6 cpm/ml. Prehybridization and hybridization are performed in sealed plastic bags at 37 °C in a volume of 0.5–1 ml of solution/10 cm² tissue surface area. After hybridization, sections are washed first in $2 \times$ SSC at 37° for 30 min, then in $2 \times$ SSC at 55° for 30 min. Following the final wash, sections are stained with hematoxylin and eosin, set against Saran wrap to dry, and exposed to film (Kodak XAR-5) at −70 °C with an intensifying screen for 24–72 h.

2.2 Protein Blotting

Sections are placed against 0.2 μm nylon membranes (Pall Biodyne) for 10–30 min for passive protein transfer. Next, the membranes with sections attached are immersed in 70% Laemmli buffer (Laemmli 1970) and 30% methanol for 15 min. They are then placed in 10% acetic acid 10% methanol for 30 min. This fixes proteins to the membranes and dissolves the tape adhesive. The tape is stripped from the membranes and the membranes are rinsed in water for 5–10 min. Tissue adherent to the membranes is stripped away with a glass slide or a razor blade. Membranes are allowed to dry and either used immediately or stored at 4 °C.

Prior to antibody incubations, nonspecific protein reactive sites on membranes are blocked by incubating membranes in Blotto (5% nonfat dry milk, 0.01% simethicone and 0.001% thimerosal in PBS) (Johnson et al. 1984) either overnight at 4 °C or 2–4 hours at room temperature. Although we have used both polyclonal and monoclonal primary antibodies for detection of viral and host proteins, we usually use polyclonal antibodies because more epitopes are available and any one epitope may not survive protein denaturation during membrane preparation. Polyclonal antibodies are often reactive with gastrointestinal flora. Incubation of antibodies with a minced preparation of rodent intestines reduces this background reactivity. The optimal concentration of antibody varies, but is usually 5- to 10-fold higher than that required for immunohistochemistry in cryostat sections. Membranes are incubated in antibody diluted in Blotto for either 4 h at room temperature or overnight at 4 °C. Membranes are rinsed in Blotto and then incubated in ^{125}I-labeled staphylococcal protein A (^{125}I-SPA) at a concentration of 5×10^5 cpm/ml for 1 h. After incubation in ^{125}I-SPA, membranes are washed first in Blotto for 30 min, then in 0.5 m LiCl 0.10 M TRIS, pH 8.0, 1% NP40 for 20 min. They are then rinsed in water, drained and set directly against a fine-grain film (XRP-1 Kodak) for 24–72 h.

2.3 Specificity and Sensitivity of WAS In Situ Hybridization and Protein Blotting

The specificity and sensitivity of macroscopic section in situ hybridization and protein blotting for detection of nucleic acids and proteins will depend on the viral infection under study and the probes and antibodies employed. The specificity of hybridization or immunoreactivity should be tested in each experiment with appropriate negative controls. These usually include sections taken from uninfected animals and from animals infected with an agent with which probes or antibodies should not react.

Figure 2 shows the application of this approach to demonstration of specificity in WAS protein blots taken from mice infected with reovirus type 3, an agent which replicates in central nervous system (CNS) tissues (Walters et al. 1963), and lymphocytic choriomeningitis virus (LCMV), an agent which causes multiorgan infection (Buchmeier et al. 1980). Antibodies to each of these viruses only react in the WAS from the animal infected with that virus. Sections from uninfected animals and animals infected with the alternate virus showed no immunoreactivity except for background signal in gastrointestinal tract (Fig. 2E, F). Background reactivity can be reduced, and in some instances eliminated, by adsorbing antibodies with minced preparations of intestinal tract or WAS from uninfected animals. Similar controls have been used to confirm specificity of hybridization to viral nucleic acids in WAS from infected animals. In studying protein blots to detect endogenous antigens, organs known not to contain the target antigens serve as controls for

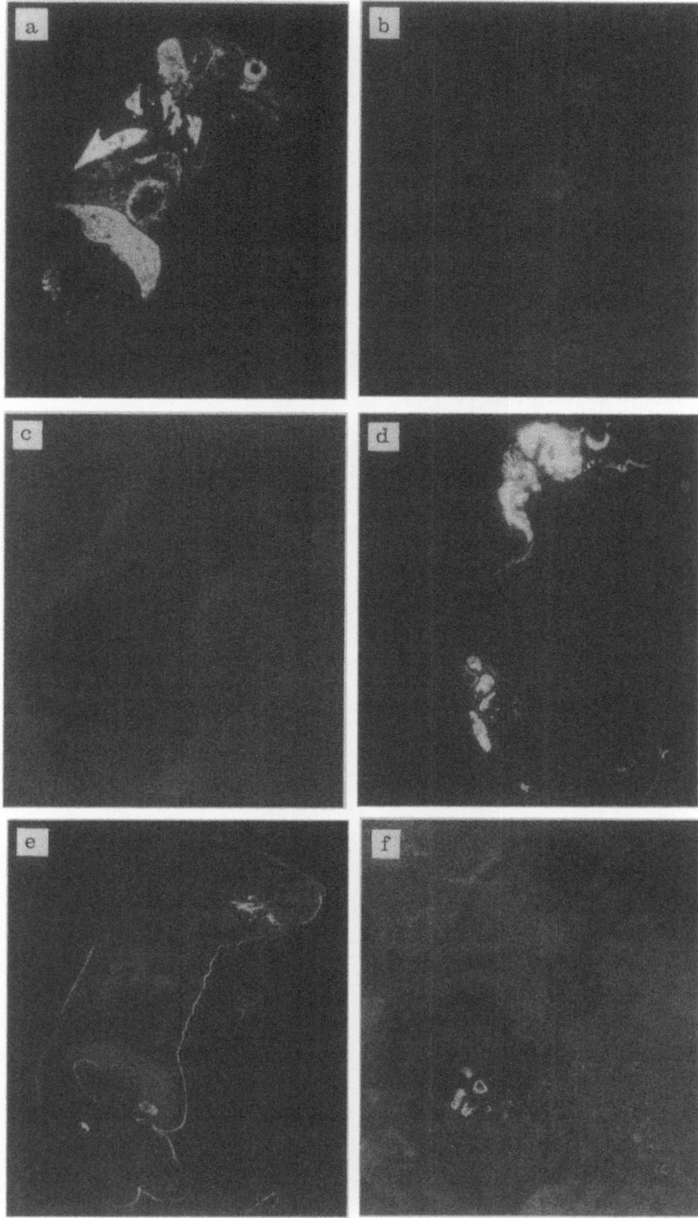

Fig. 2a–f. Specificity of antigen detection in protein blotting. **a–c** Sagittal mouse sections incubated with guinea pig antibody to lymphocytic choriomeningitis virus (LCMV); **d–f** sections incubated with rabbit antiserum to reovirus (REOV). Sections in **a** and **e** were taken from a mouse infected with LCMV; sections in **b** and **d** were taken from a mouse infected with reovirus (REOV); sections in **c** and **f** were from uninfected mice. (From Lipkin and Oldstone 1986)

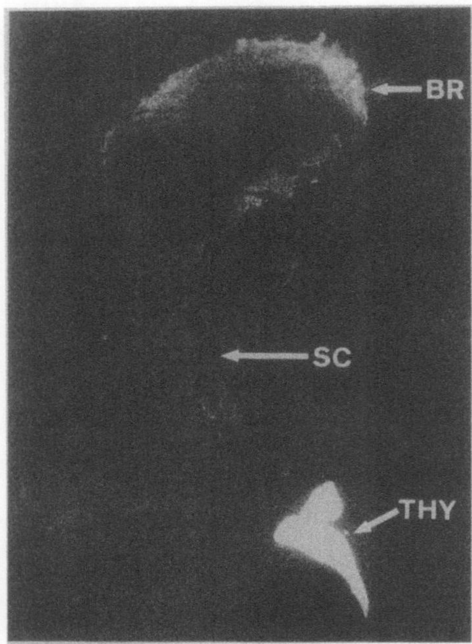

Fig. 3. Recognition of endogenous iso-antigens in a mouse protein blot. Sagittal mouse section was incubated with rabbit antibody to Thy 1.2 and ^{125}I-labeled staphylococcal protein A. *BR*, brain; *SC*, spinal cord; *THY*, thymus. (From LIPKIN and OLDSTONE 1986)

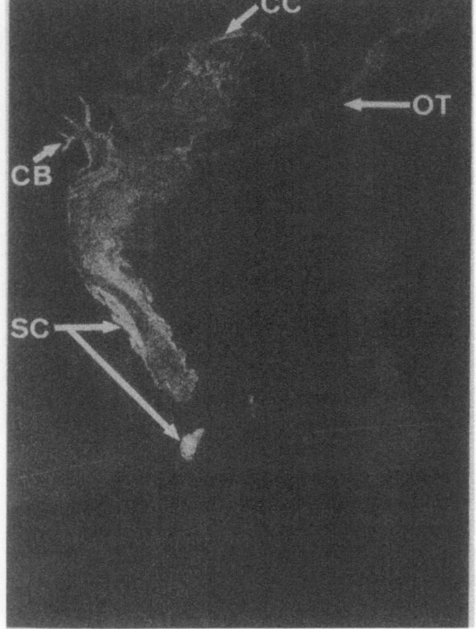

Fig. 4. Distribution of myelin basic protein in a mouse protein blot. Sagittal mouse section was incubated with rabbit antibody to MBP and ^{125}I-labeled staphylococcal protein A. *CB*, cerebellum; *CC*, corpus callosum; *OT*, olfactory tract; *SC*, spinal cord. (From LIPKIN and OLDSTONE 1986)

the specificity of the antibody. Figures 3 and 4 illustrate the sensitivity, resolution and specificity of protein blotting for detection of Θ isoantigen and myelin basic protein, respectively. Antibody to Θ isoantigen labels CNS tissues, including brain and spinal cord as well as thymus (Fig. 3). Antibody to myelin basic protein also labels CNS tissues, but the distribution of signal is confined to white matter structures including the corpus callosum, olfactory tract, and myelinated fibers in cerebellum and spinal cord (Fig. 4).

In the LCMV system, the threshold for antigen detection was determined by placing LCMV proteins onto membranes, allowing passive transfer, and then processing membranes using the protein blotting protocol described. These experiments allowed us to detect LCMV antigen concentrations as low as 10- to 20-ng/cm² membrane surface area (LIPKIN and OLDSTONE 1986). The sensitivity of WAS in situ hybridization was examined by processing 40-μm sections from LCMV-infected animals either for in situ hybridization as described or for nucleic acid extraction with sodium dodecyl sulfate proteinase K and phenol and dot blot hybridization. In these experiments, WAS in situ hybridization was 10%–20% as efficient as cytoplasmic RNA dot blots for detection of LCMV nucleic acids (SOUTHERN et al. 1984).

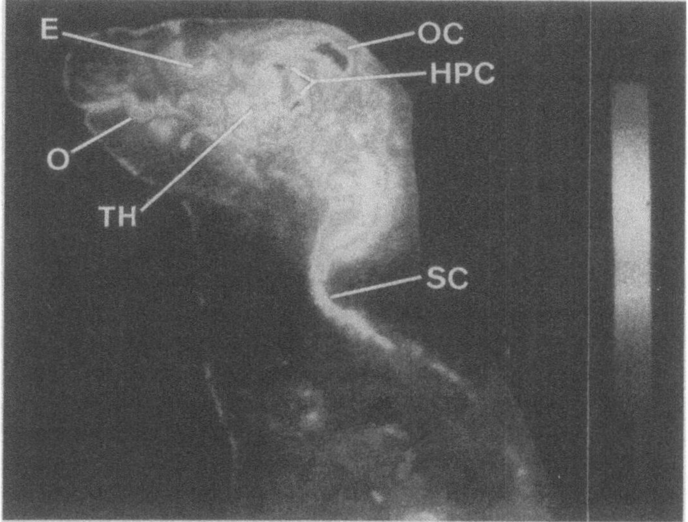

Fig. 5. Quantitative distribution of reovirus (REOV) antigens in a mouse acutely infected with REOV. Sagittal section incubated with rabbit antibody to REOV and ¹²⁵I-labeled staphylococcal protein A. Autoradiogram was scanned with a laser densitometer for color-coded densitometry. *E*, eye (retina); *HPC*, hippocampus; *O*, oropharynx; *OC*, occipital cortex; *SC*, spinal cord. REOV antigen concentration is indicated by color: red is highest, yellow is intermediate, blue is lowest (color bar at *right* of figure)
(see also Appendix for color illustration)

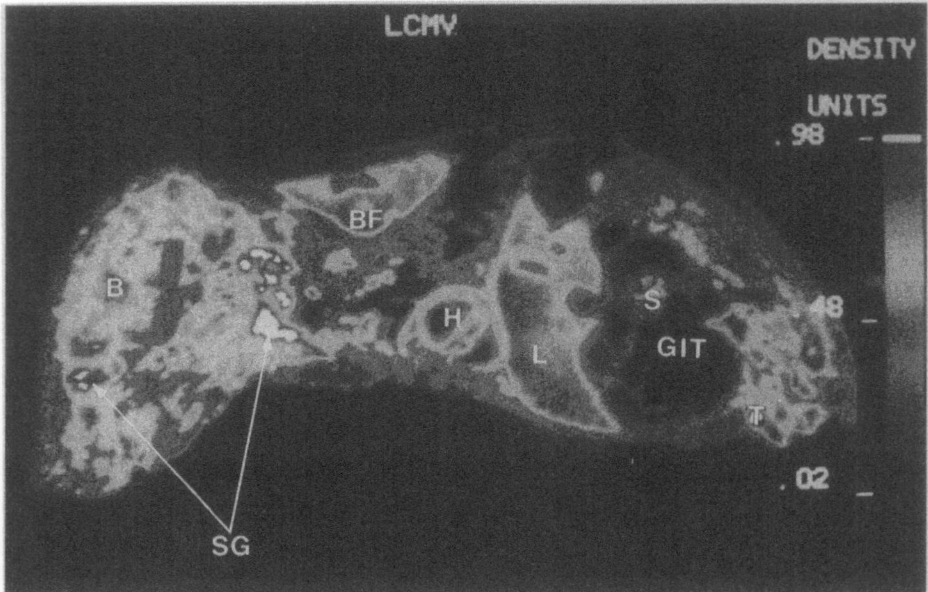

Fig. 6. Quantitative distribution of lymphocytic choriomeningitis virus (LCMV) antigens in a mouse persistently infected with LCMV. Sagittal mouse section incubated with guinea pig antibody to LCMV and ^{125}I-labeled staphylococcal protein A. Autoradiograph was scanned with a laser densitometer for color-coded densitometry. *B*, brain; *BF*, brown fat; *GIT*, gastrointestinal tract; *H*, heart; *L*, liver; *S*, spleen; *SG*, salivary glands; *T*, testes. LCMV antigen concentration is indicated by color: white is highest with decreasing concentrations shown in red, yellow, green and blue (color bar at *right* of figure)
(see also Appendix for color illustration)

2.4 Quantitation of Antigens and Nucleic Acids in WAS

Autoradiographs generated through in situ hybridization or protein blotting can be analyzed via laser densitometry to quantitate nucleic acid or protein concentration by anatomic region. Application of laser densitometry to analysis of a reovirus (REOV) protein blot is shown in Fig. 5. Regions with the highest signal intensity (occipital cortex, thalamus and hippocampus) are indicated in red; lower signal intensity is present as yellow in retina, remainder of brain, spinal cord and oropharynx. Laser densitometry can be particularly useful for comparing concentrations of antigens in widely disparate locations. Figure 6 is a density map of viral antigens in a mouse persistently infected with LCMV. Highest concentrations of LCMV antigens are shown in white (salivary glands, testes), lower concentrations in red and yellow (brain, brown fat, liver, spleen); still lower concentrations are blue (gastrointestinal tract).

3 Applications of the WAS Technique to Molecular Analyses of Viral Pathogenisis

3.1 Studies of the Distribution of Viral Antigens and Nucleic Acids

WAS simultaneously samples multiple tissues and thus serves as a convenient tool for mapping the location of viral genes and proteins in infected animals and assessing changes in these maps over the course of infection or following experimental manipulation. In some cases, WAS analysis can lead to identification of novel distributions of target nucleic acids and proteins, e.g., WAS protein blots of mice infected with reovirus have shown high levels of viral antigens in the oropharynx. Prior to these studies the oropharynx had not been appreciated as a site for replication of reovirus (LIPKIN and OLDSTONE 1986). WAS can also provide insight into the effects of an infection on the expression of endogenous (host) genes and gene products. Table 2 lists endogenous genes and gene products that have been identified in WAS.

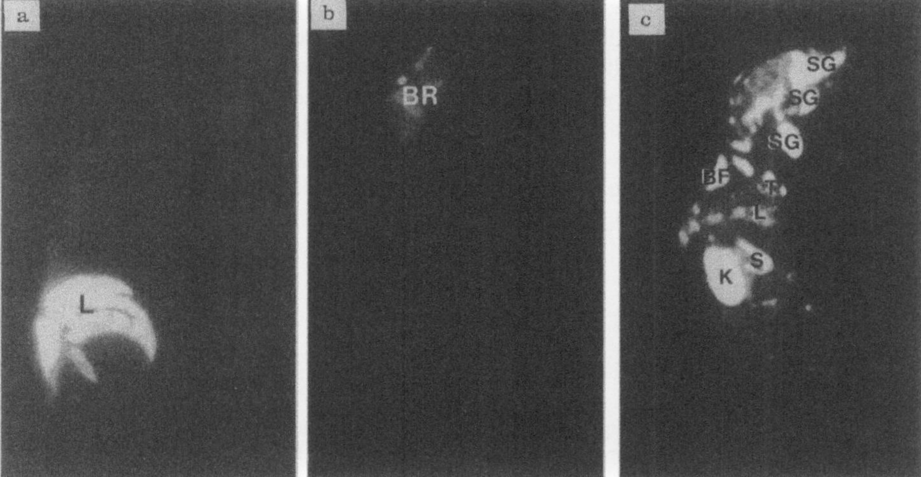

Fig. 7a–c. Distribution of viral nucleic acids in rodents infected with ground squirrel hepatitis virus (GSHV), rabies virus (RV) or LCMV. Hybridization to liver is seen in the chipmunk infected with GSHV (**a**), to brain in the mouse infected with RV (**b**) and to multiple organs in the mouse infected with LCMV (**c**), using ^{32}P-labeled cDNA probes to GSHV, RV and LCMV, respectively. The specificity of each of these probes for its target viral sequence(s) was confirmed in hybridization experiments with WAS from uninfected animals and animals infected with alternate viruses. *BF*, brown fat; *BR*, brain; *K*, kidney; *L*, liver; *SG*, salivary glands; *T*, thymus

Table 2. Endogeneous host proteins detected in macroscopic sections

---

Glial fibrillary acidic protein (BLOUNT et al. 1986)

Growth hormone (BLOUNT et al. 1986)

Myelin basic protein (LIPKIN and OLDSTONE 1986; BLOUNT et al. 1986)

Prolactin (BLOUNT et al. 1986)

θ-Isoantigen (LIPKIN and OLDSTONE 1986; BLOUNT et al. 1986)

The ability to display genes and gene products in an anatomic context allows rapid comparisons of tropism with different viral agents. Figure 7 shows the differences in distribution of viral nucleic acids in chipmunks infected with ground squirrel hepatitis virus (GSHV), mice infected with rabies virus, and mice infected with LCMV. Nucleic acid hybridization signal is confined to the liver in chipmunks infected with GSHV (Fig. 7A) and to the CNS in mice infected with RV (Fig. 7B). In contrast, animals inoculated at birth with LCMV have persistent infection in most tissues. The catholic nature of this infection is indicated by the presence of LCMV nucleic acids in multiple organs including brown fat, kidney, liver, salivary glands and spleen (Fig. 7C).

WAS analysis has proved invaluable too in studying the pathogenesis of polyomavirus (PV) infections (Fig. 8). Its use has identified the distribution patterns of viral nucleic acid over the course of infection and has established which organs are infected during primary, acute and persistent phases of infection (DUBENSKY et al., 1984, DUBENSKY and VILLARREAL 1984). In addition, it has been employed to determine the effects of the different routes of infection on viral tropism and to dissect the molecular genetic basis of PyV tropism to tissues through studies with PV recombinants. Figure 8, column 1 shows the primary replication

Fig. 8. In vivo tissue specificity of polyomavirus (PV) infection. Whole mouse hybridization ▶ patterns in sagittal sections of newborn mice inoculated with PV, hybridized with ^{32}P-labeld cDNA probe to PV. The *left column* shows both the primary site of replication in the serosal surfaces and the acute/secondary sites of replication after intraperitoneal inoculation with the wild-type strain PyA2. A stained tissue section is included as a reference. In the *middle column*, newborn mice are inoculated intranasally with PyA2. The *right column* demonstrates how changes in the enhancer region affect tissue specificity. The MoMuLV/Py enhancer recombinant contains the MoMuLV enhancer in place of the B enhancer region of PV. The LPT/A2 recombinant contains the encoding region of the Large Plaque Toronto strain of PV in the coding background of PyA2. The mixed enhancer stock contains several different enhancer types. Contact prints were made of autoradiograms. Organs are labeled as follows: *C*, clavicle (bone marrow); *F*, fat pad; *H*, heart; *K*, kidney; *L*, lung; *P*, pancreas; *Pg*, parotid gland; *S*, spleen; *T*, thymus

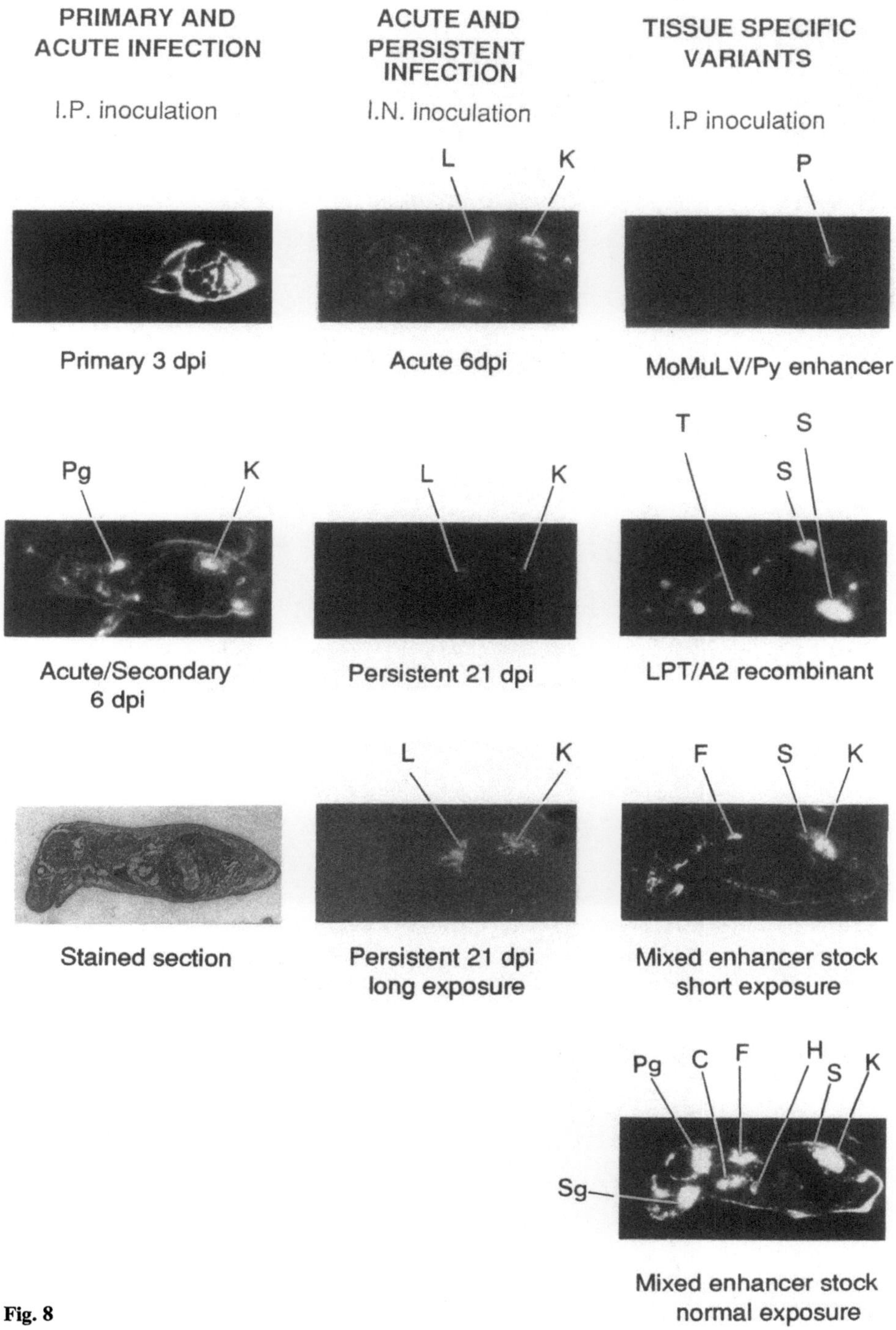
PRIMARY AND ACUTE INFECTION
I.P. inoculation
ACUTE AND PERSISTENT INFECTION
I.N. inoculation
TISSUE SPECIFIC VARIANTS
I.P inoculation
L
K
P
Primary 3 dpi
Acute 6dpi
MoMuLV/Py enhancer
Pg
K
L
K
T
S
S
S
Acute/Secondary 6 dpi
Persistent 21 dpi
LPT/A2 recombinant
L
K
F
S
K
Stained section
Persistent 21 dpi long exposure
Mixed enhancer stock short exposure
Pg
C
F
H
S
K
Sg
Mixed enhancer stock normal exposure

Fig. 8

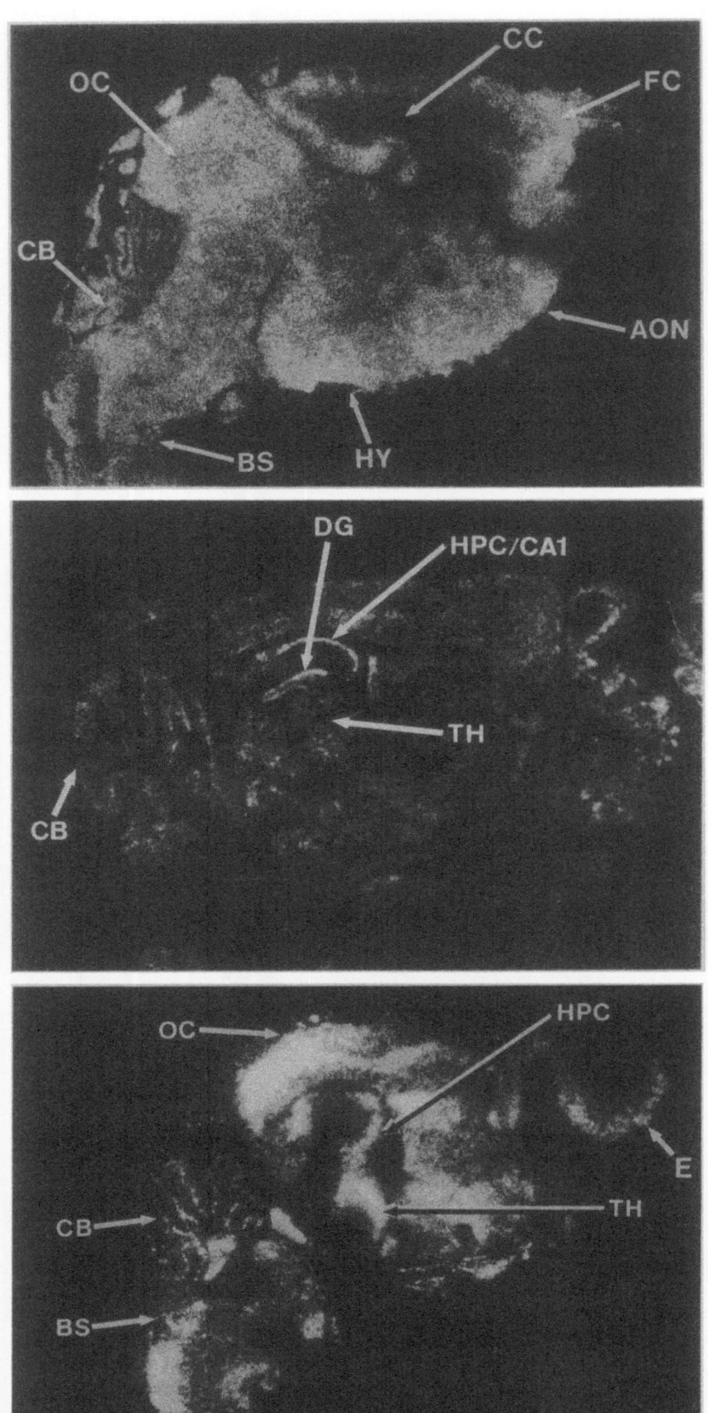

Fig. 9a–c

in serosal tissue followed by secondary replication predominantly in parotid gland and kidney after intraperitoneal inoculation of newborn mice with wild-type PyA2. In contrast, intranasal inoculation of mice leads to replication predominantly in the lungs along with kidney (Fig. 8, column 2). Thus, rather different organ tropism can result from different routes of infection with Py and indicate that there are some clear restrictions on virus accessibility to otherwise fully permissive organs. The route of inoculation also affects the site of persistent infection (Fig. 8, column 2), as persistent lung infection only occurs following intranasal inoculation. An important observation that was first made using WAS in situ hybridization with PyV infection was how changes in noncoding regulatory DNA can have dramatic effects on in vivo tissue specificity. When the β-enhancer region is replaced by the moloney murine leukemia virus enhancer DNA (MoMuLV), the resulting polyoma virus recombinant infects the pancreas almost exclusively (ROCHFORD et al. 1987). As this is a most unexpected site for PyV replication, it is likely that this observation would not have been made without WAS in situ hybridization. The rather dramatic effects of other enhancer changes on tissue tropism are also shown (Fig. 8, column 3). Exchanging the enhancer from the Large Plaque Toronto strain into the PyA2 genome results in a virus which can replicate in thymus and spleen as well as other sites. Other enhancer changes can lead to infection of additional tissues such as that seen following infection with PyV containing a multermerized enhancer from PyA2 and MoMuLV. Here (Fig. 8, column 3), viral DNA is replicated in bone marrow, brown fat, heart (atrium), kidney, salivary gland and spleen (ROCHFORD-SUN and VILLARREAL, unpublished material). Thus, it appears that PV tissue specificity is not completely dependent on the specific interaction of its capsid with cellular viral receptors, but is rather dependent on the nature of its regulatory DNA.

While the resolution in ^{32}P in situ autoradiographs is limited to the whole organ level, in ^{125}I-labeled protein blots the signal is sufficiently discrete to map viral antigens within organs. This high resolution of protein blots has been exploited to map the location of viral antigens to tracts and nuclei within the CNS in mice infected with rabies virus, LCMV and reovirus Mice infected with rabies virus have diffuse, homogeneous expression of viral antigens in the CNS (Fig. 9A). The expression of viral antigens in mice infected with LCMV is also present throughout the CNS but is more highly concentrated in dentate gyrus and sector CA1 of hippocampus (Fig. 9B). A third pattern of viral antigen distribution is found in mice infected with reovirus. Reovirus antigen is again present throughout the CNS but with highest concentration in hippocampus, occipital cortex, retina and thalamus (Fig. 9C).

Protein blot analysis has been extended in the reovirus system to address the effects on viral tropism of monoclonal antibody driven mutations. Figure 10 shows

◀ **Fig. 9a–c.** Distribution of viral antigens in brains of mice infected with rabies virus (RV), lymphocytic choriomeningitis virus (LCMV) or reovirus (REOV). Sagittal mouse sections were incubated with rabbit antibody to RV **(a)**, guinea pig antibody to LCMV, **(b)** or rabbit antibody to REOV **(c)** and ^{125}I-labeled staphylococcal protein A. *AON*, anterior olfactory nucleus; *BS*, brainstem; *CB*, cerebellum; *CC*, corpus callosum; *DG*, dentate gyrus; *E*, eye (retina); *FC*, frontal cortex; *HPC/CA1*, sector CA1 of hippocampus; *HY*, hypothalamus; *OC*, occipital cortex; *TH*, thalamus

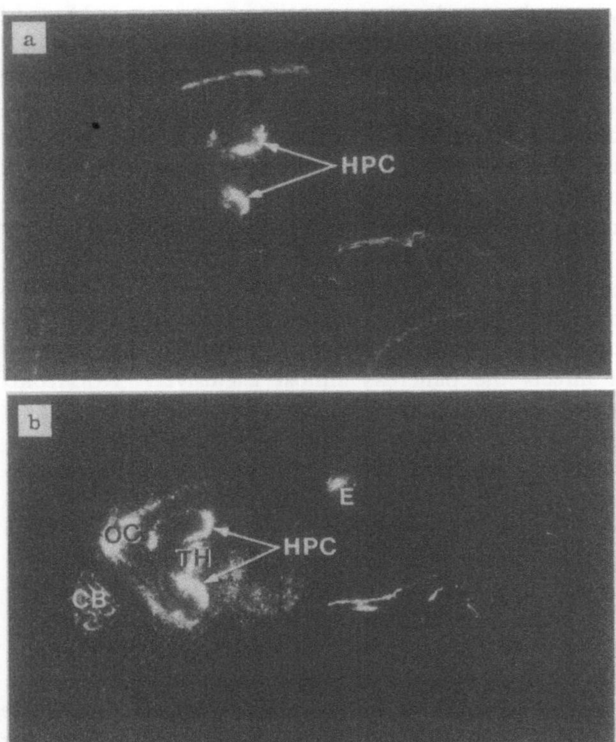

Fig. 10a, b. Distribution of viral antigens in brains of mice acutely infected with **a** monoclonal antibody selected variant reovirus (REOV-K) or **b** wild-type reovirus (REOV). Sagittal sections incubated with rabbit antibody to REOV and ^{125}I-labeled staphylococcal protein A. *CB*, cerebellum; *E*, eye (retina); *HPC*, hippocampus; *OC*, occipital cortex; *TH*, thalamus

application of WAS protein blots to study sites of viral antigen expression in mice infected with either wild-type reovirus or the monoclonal antibody selected variant strain, reovirus-K. As before (Fig. 9c), the sites of viral antigen expression in reovirus-infected mice are highest in hippocampus, occipital cortex, retina and the thalamus (Fig. 10b). In mice infected with reovirus-K, however, viral antigens are confined to the hippocampus (Fig. 10a). This difference in tropism has been confirmed independently with immunohistochemistry in cryostat sections (SPRIGGS et al. 1983).

3.2 Viral Persistence

WAS have been particularly useful to the study of viral persistence in the LCMV murine system. In situ hybridization experiments with WAS led to recognition of two phenomena that may be important to the mechanism of viral persistence in vivo (SOUTHERN et al. 1984): first, that the accumulation of viral nucleic acids and

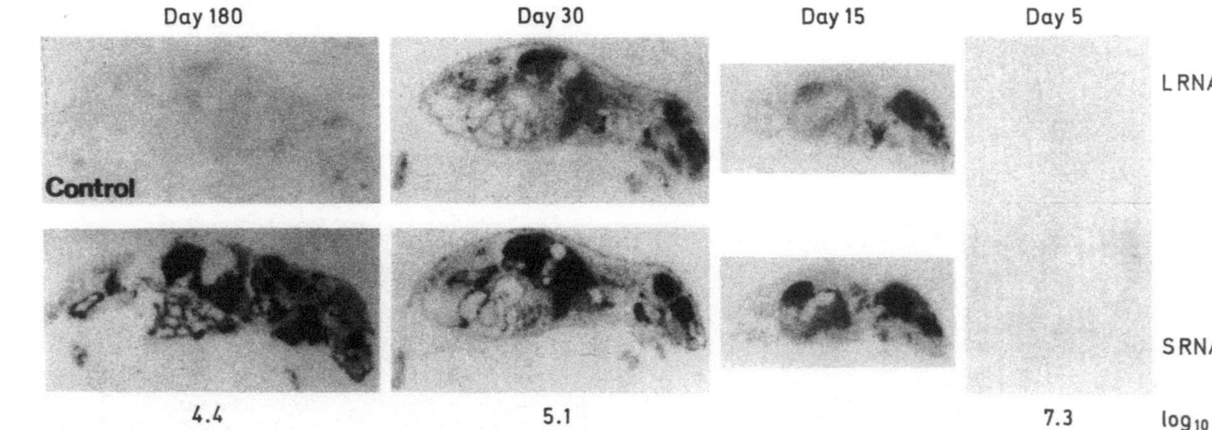

Fig. 11. Distribution of L and S segment viral nucleic acids in SWR/J mice persistently infected with LCMV. In situ hybridization with ^{32}P-labeled probes to either the L or the S segment of the LCMV genome in mice of different ages, inoculated at birth with LCMV. Titers of infectious virus per gram of brain in $\log_{10}$ are displayed for each time point (SOUTHERN et al. 1984)

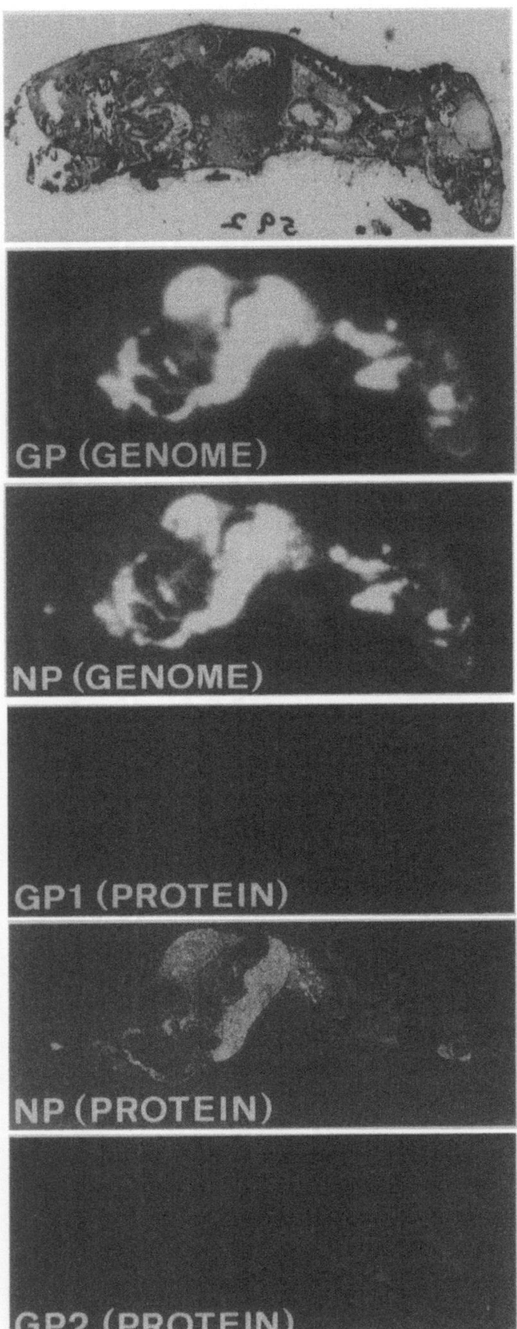

Fig. 12. Distribution of viral S segment transcripts and protein products in a 2-month old mouse persistently infected with LCMV. WAS have been hybridized with ^{32}P-labeled strand-specific probes to detect nucleoprotein (NP) or glycoprotein (GP) mRNAs, or incubated with monoclonal antibodies and ^{125}I-labeled staphylococcal protein A to detect NP, GP-1 or GP-2. The cRNA probe to GP mRNA is complementary to both genomic RNA and GP mRNA. A WAS stained with hematoxylin and eosin is included as a reference (unpublished data, BLOUNT and OLDSTONE; biochemical data, FULLER-PACE and SOUTHERN 1988)

of infectious virus are dissociated, and second, that one segment of the bisegmented LCMV genome accumulates more than the other. Figure 11 illustrates these points. Mice inoculated at birth with LCMV were sacrificed 5, 15, 30 or 180 days later for WAS in situ hybridization with probes to either the L or the S segment of LCMV genome. Brains from persistently infected littermates were taken at the same time points for viral plaque assays. In spite of increasing concentrations of target L and S segment viral nucleic acids, titers of infectious virus decreased nearly 3 logs. This suggested the possibility that persistently infected mice generated incomplete viral particles. Such incomplete virus particles have been further defined for LCMV as being defective and subgenomic (WELSH and BUCHMEIER 1979; FRANCIS and SOUTHERN 1988) and have been postulated to be important for persistence of viruses (PERRAULT 1981; LAZZARINI et al. 1981; RAO and HUANG 1982). In addition, WAS studies revealed that S segment nucleic acid sequences accumulate in excess of L segment sequences. The significance of this finding is unclear, but it may be related to maintenance of persistence, since the L segment encodes the LCMV polymerase (SINGH et al. 1987).

The level of expression of individual RNAs and proteins encoded by the S segment of LCMV have been analyzed in WAS during persistent infection. Expression of viral glycoprotein (GP) is reduced relative to expression of viral nucleoprotein (NP) in vitro (Welsh and Buchmeier 1979) and in vivo (OLDSTONE and BUCHMEIER 1982; RODRIGUEZ et al. 1983). Reduced GP on the surface of infected cells may contribute to the phenomenon of persistence by allowing these cells to evade immune surveillance (OLDSTONE and BUCHMEIER 1982; RIVIERE et al. 1986). In persistent LCMV infection in vitro, levels of GP mRNA are not significantly reduced relative to levels of NP mRNA (FRANCIS and SOUTHERN 1988), indicating that the basis for decreased expression of GP is posttranscriptional. Figure 12 shows five WAS from one mouse persistently infected with LCMV, hybridized with strand-specific cRNA probes to detect GP mRNA, or incubated with antibodies to detect NP, GP-1 or GP-2. These studies, performed a few years prior to the in vitro analysis, demonstrated GP mRNA and NP expression in multiple tissues. GP-1 and GP-2 were not visualized.

Unlike neonatal mice, which become persistently infected following inoculation with LCMV, mice infected as adults mount a LCMV-specific cytotoxic T-lymphocyte (CTL) response and clear the virus. An exception to this pattern is seen when adult mice are inoculated with a viral variant (clone 13) isolated from the lymphoid tissues (spleen) of a mouse persistently infected with LCMV (AHMED et al. 1984; OLDSTONE et al. 1988). Adult mice inoculated with clone 13 become persistently infected. Failure to clear virus is associated with viral infection of lymphocytes and with a decrease in numbers of circulating lymphocytes. These data suggest that the variant may cause persistent infection in adult mice by destroying lymphocytes (OLDSTONE et al. 1988). Recently, direct RNA sequencing of the clone 13 lymphocyte variant and the parental strain of LCMV (Armstrong 53B) has mapped the basis for the clone 13 phenotype to a region of the L RNA segment likely encoding the viral polymerase, and not to the S RNA segment (Salvato et al. 1988, and unpublished data). Figure 13 shows results of WAS in situ hybridization experiments in adult mice inoculated with either clone 13 or Armstrong 53B brain isolate. Two months after infection, mice infected with clone 13 had LCMV nucleic

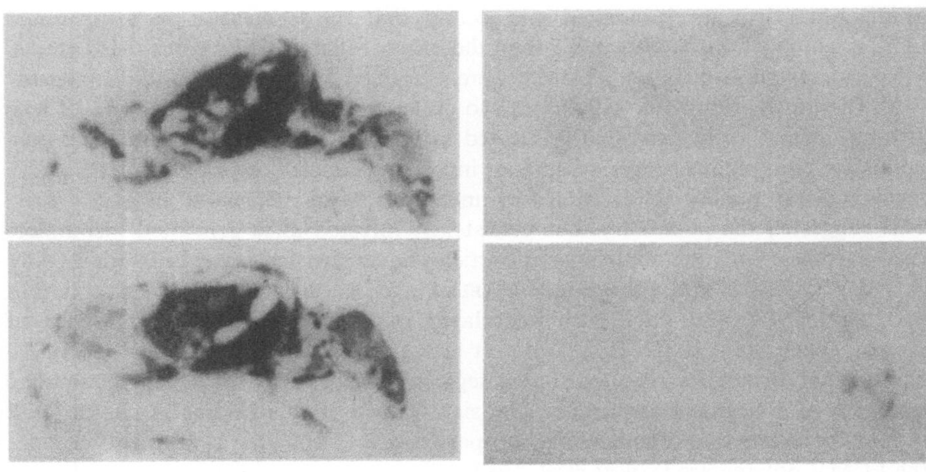

Fig. 13. Distribution of viral nucleic acids in mice infected as adults with either the Armstrong 53 B (brain isolate; parental) strain of LCMV or an LCMV variant (lymphoid isolate — Clone 13) isolated from the spleen of a mouse persistently infected with Armstrong 53 B. Six-week-old mice were inoculated intravenously with $1–2 \times 10^8$ plaque-forming units of clone 13 or Armstrong 53 B. Two months later, animals were sacrified for infectious center assays and WAS in situ hybridization with ^{32}P-labeled probes to detect LCMV nucleic acids. *CTL*, cytotoxic T-lymphocyte

acids in multiple tissues. No LCMV nucleic acids were seen in the animals infected with Armstrong 53B because CTL had been generated and virus had been cleared.

In addition to studies directed toward understanding the biology of the virus in LCMV infection in vivo, WAS have been used to explore the role of the host response in persistent infection. Although antibodies to LCMV proteins are generated during persistent infection, LCMV-specific CTL responses are selectively decreased (AHMED et al. 1984). To investigate the role of CTL in persistent infection, mice persistently infected with LCMV were transfused with LCMV-immune splenocytes and assayed for clearance of viral nucleic acids and proteins using WAS. These experiments confirmed that clearance of LCMV in this system was by CTL, i.e., virus-specific H-2 restricted Th 1.2^+ CD8$^+$ (LYT2$^+$) CD4^{nil} (L3T4) lymphocytes. Further, kinetic studies following adoptive transfer identified two sites, kidney and CNS, which were relatively refractory to viral clearance. This led to recognition of a novel mechanism for viral clearance from the CNS (OLDSTONE et al. 1986).

Mice were sacrified 15, 30, 60 or 120 days after adoptive transfer for WAS analysis and viral titer studies. Fifteen days after adoptive transfer, LCMV nucleic acid sequences were cleared from liver, spleen and lungs. Brain and kidney contained LCMV nucleic acids at 15 and 30 days after transfer, only clearing between days 30 and 120 after transfer (Fig. 14). WAS protein blots showed the same kinetics for LCMV antigen clearance (Fig. 15). These results, coupled with further studies using cryostat sections, identified the mechanism for clearance of viral products in different tissues in the adoptive transfer model. Liver, lung and kidney showed parenchymal infiltration with inflammatory cells and necrosis, the result of CTL recognition and lysis. The continued expression of LCMV products in kidney was found to represent trapping of immune complexes in the glomeruli

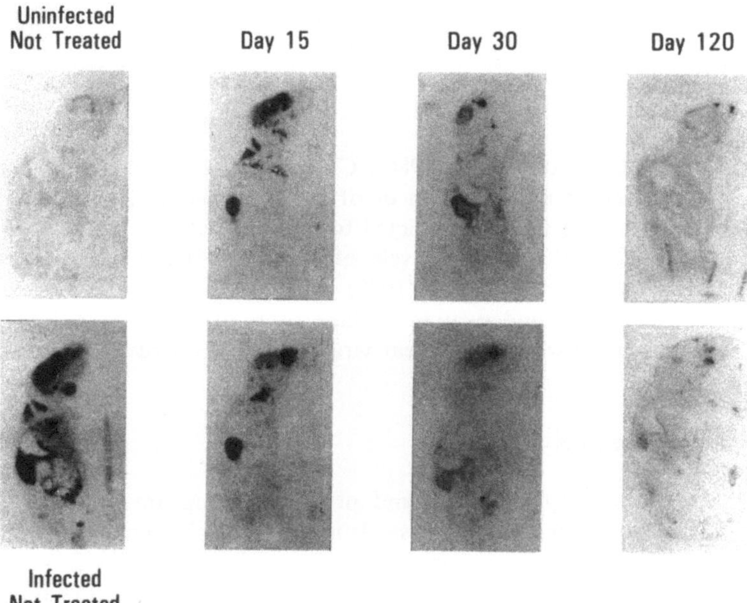

Tissue	LCMV titer ($\log_{10}$ PFU/organ or ml)	
	PI Untreated	120 days Post Transfer
Serum	4.5	<1.6
Spleen	5.8	<1.6
Liver	5.3	<1.6
Lung	5.3	<1.6
Brain	5.2	1.9
Kidney	6.5	3.9

Fig. 14. Clearance of infectious virus and viral nucleic acid sequences by viral specific H-2 restricted immune splenocytes bearing the phenotype of Th 1.2^+ CD8$^+$ CD4^{nil} (CTL) in mice persistently infected with LCMV. Mice persistently infected with LCMV were transfused with CTL obtained from mice immunized with LCMV. Transfused mice were sacrificed 15, 30 or 120 days later either for WAS in situ hybridization analysis with ^{32}P-labeled probes to LCMV nucleic acids or for viral titer studies. See OLDSTONE et al. 1986 for methodology

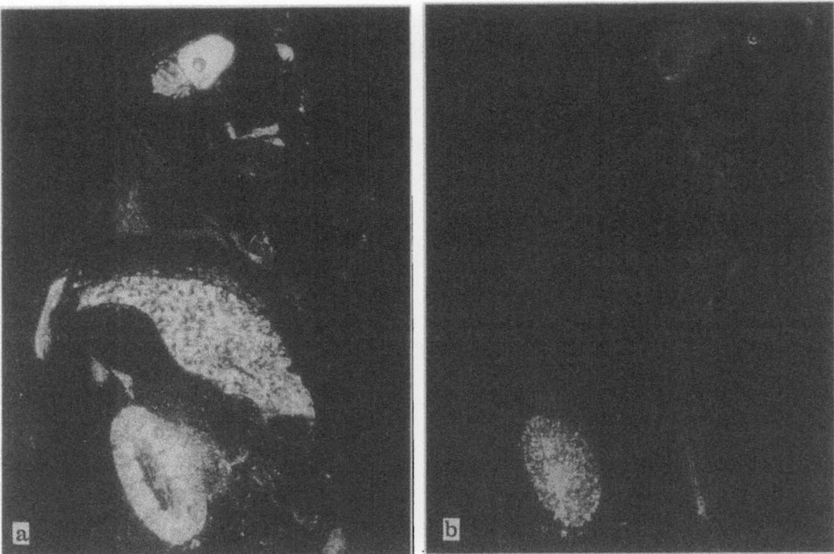

Fig. 15a, b. Clearance of viral antigens by viral immune splenocytes in a mouse persistently infected with LCMV. See legend for Fig. 14. LCMV antigens are seen in multiple organs in a 4-month-old persistently infected mouse (a). b Distribution of LCMV antigens in 4-month-old persistently infected mouse 60 days after adoptive transfer of LCMV immune splenocytes. Sagittal section incubated with guinea pig antibody to LCMV and ^{125}I-labeled staphylococcal protein A

(Fig 15; OLDSTONE et al. 1986). CNS tissue taken at several time points showed no evidence of inflammation or of neuronal damage, presumably because LCMV infection in the CNS is restricted to neurons (RODRIGUEZ et al. 1983) and neurons do not express sufficient levels of class I MHC products to be recognized by CTL (VITETTA and CAPRA 1978; WONG et al. 1984; KLAVINSKIS et al. 1988). Experiments are underway in the laboratory of one of us (MBAO) to determine the mechanism by which CTL clear virus from CNS tissues without neuronal lysis.

4 Conclusions

WAS in situ hybridization and protein blotting are powerful experimental tools for studies in viral pathogenesis. It is now possible to map the anatomic distribution of host and viral genes, messages and proteins in whole animals during natural and experimental infections and to examine the effects of infections on distribution of host genes and proteins. The macroscopic view obtained through WAS includes multiple regions and tissues, some of which might not be selected a priori. This affords the investigator an opportunity for novel observations even in well described systems, and suggests application to analysis of transgenic animal models. WAS will also be of value in addressing basic questions regarding tissue expression of transcripts and proteins in developmental biology.

The authors wish to acknowledge the following investigators for reagents used in these studies: Bernard Fields, Donald Ganem, Ken Tyler, Harold Varmus and William Wunner. We also thank Diane Nolin for her efforts in preparation of the manuscript.

References

Ahmed R, Salmi A, Butler LD, Chiller JM, Oldstone MBA (1984) Selection of genetic variants of lymphocytic choriomeningitis virus in spleens of persistently infected mice: role in suppression of cytotoxic T lymphocyte response and viral persistence. J Exp Med 60: 521–540

Blount P, Elder J, Lipkin WI, Southern PJ, Buchmeier MJ, Oldstone MBA (1986) Dissecting the molecular anatomy of the nervous system: analysis of RNA and protein expression in whole body sections of laboratory animals. Brain Res 382: 257–265

Brahic M, Stowring L, Ventura P, Haase AT (1981) Gene expression in visna virus infection. Nature 292: 240–242

Buchmeier MJ, Welsh RM, Dutko FJ, Oldstone MBA (1980) The virology and immunobiology of lymphocytic choriomeningitis virus infection. Adv Immunol 30: 275–331

Dubensky TW, Villarreal LP (1984) The primary site of replication alters the evenual site of persistent infection by polyoma virus in mice. J Virol 50: 541–546

Dubensky TW, Murphy FA, Villarreal LP (1984) The detection of DNA and RNA virus genomes in the organ systems of whole mice: patterns of mouse organ infection by polyoma virus. J Virol 50: 779—783

Feinberg AP, Vogelstein BA (1983) A technique for radiolabeling DNA restriction endonuclease fragments to high specific activity. Anal Biochem 132: 6–13

Francis SJ, Southern PJ (1988) Molecular analysis of viral RNAs in mice persistently infected with lymphocytic choriomeningitis virus. J Virol 62: 1251–1257

Fuller-Pace FV, Southern PJ (1988) Temporal analysis of transcription and replication during acute infection with lymphocytic choriomeningitis virus. Virology 162: 260–263

Haase AT, Stowring L, Harris JD, Traynor B, Ventura P, Peluso P, Brahic M (1982) Visna DNA synthesis and the tempo of infection in vitro. Virology 119: 399–410

Haase AT, Gantz D, Eble B, Walker D, Stowring R, Ventura Ap, Blum H, Wietgrefe S, Zupancic M, Tourtellote W, Gibbs CJ, Norrby E, Rozenblatt S (1985a) Natural history of restricted synthesis and expression of measles virus genes in subacute sclerosing panencephalitis. Proc. Natl. Acad Sci USA 82: 3020–3024

Haase AT, Gantz D, Blum H, Stowring L, Ventura P, Geballe A, Moyer B, Brahic M (1985b) Combined macroscopic and microscopic detection of viral genes in tissues. Virology 140: 201–260

Johnson DA, Gautsch JW, Sportsman JR, Elder JH (1984) Improved techniques utilizing non-fat dry milk for analysis of protein and nucleic acids transferred to nitrocellulose. Gene Analysis Techniques 1: 3–8

Klavinskis LS, Tishon A, Oldstone MBA (manuscript in preparation)

Laemmli UK (1970) Cleavage of structural proteins during the assembly of the head of bacteriophage T4. Nature 227: 680–685

Lazzarini R, Keene JD, Schubert M (1981) The origins of defective interfering particles of the negative strand RNA viruses. Cell 26: 145–154

Lipkin WI, Oldstone MBA (1986) Analysis of endogenous and exogenous antigens in the nervous system using whole animal sections. J Neuroimmunol 11: 251–257

Maniatis T, Fritsch Ef, Sambrook J (1982) Molecular cloning. New York, Cold Spring Harbour

Oldstone MBA, Buchmeier MJ (1982) Restricted expression of viral glycoprotein in cells of persistently infected mice. Nature 300: 360–362

Oldstone MBA, Blount P, Southern PJ, Lampert PW (1986) Cytoimmunotherapy for persistent virus infection: unique clearance pattern from the central nervous system. Nature 321: 239–243

Oldstone MBA, Salvato M, Tishon A, Lewicki H (1988) Virus-lymphocyte interactions. III. Biologic parameters of a virus variant that fails to generate CTL and establishes persistent infection in immunocompetent hosts. Virology 164: 507–516

Perrault J (1981) Origin and replication of defective interfering particles. In: Shatkin AJ (ed) Initiation signals in viral gene expression. Springer, Berlin Heidelberg New York, pp 151–207 (Current topics in microbiology and immunology, vol 93)

Rao DD, Huang AS (1982) Interference among defective interfering particles of vesicular stomatitis virus. J Virol 41: 210–221

Rigby PWJ, Dieckmann M, Rhodes C, Berg P (1977) Labeling deoxyribonucleic acid to high specific activity in vitro by nick translation with DNA polymerase I. J Mol Biol 113: 231–251

Riviere Y, Southern PJ, Ahmed R, Oldstone MBA (1986) Biology of cloned cytotoxic T lymphocytes specific for lymphocytic choriomeningitis virus. V. Recognition is restricted to gene products encoded by the viral S RNA segment. J Immunol 136: 304–307

Rochford R, Campbell BA, Villarreal LP (1987) A pancreas specificity results from the combination of polyomavirus and moloney murine leukemia virus enhancer. Proc Natl Acad Sci USA 84: 449–453

Rodriguez M, Buchmeier MJ, Oldstone MBA, Lampert PW (1983) Ultrastructural localization of viral antigens in the CNS of mice persistently infected with lymphocytic choriomeningitis virus (LCMV). Am J Pathol 110: 95–110

Salvato M, Shimomaye E, Southern P, Oldstone MBA (1988) Virus-lymphocyte interactions. IV. Molecular characterization of LCMV Armstrong (CTL$^+$) and that of its variant. Clone 13 (CTL$^-$). Virology 164: 517–522

Singh M, Fuller-Pace FV, Buchmeier MJ, Southern PS (1987) Analysis of the genomic L RNA segment from lymphocytic choriomeningitis virus. Virology 161: 445–456

Southern PJ, Blunt P, Oldstone MBA (1984) Analysis of persistent virus infections by in situ hybridization to whole-mouse sections. Nature 312: 555–558

Spriggs DR, Bronson RT, Fields BN (1983) Hemagglutinin variants of reovirus type 3 have altered central nervous system tropism. Science 220: 505–507

Vitetta ES, Capra JD (1978) The protein products of the murine 17th chromosome: genetics and structure. Adv Immunol 26: 147–193

Walters MN, Joske RA, Leak PJ, Stanley NF (1963) Murine infection with reovirus. I. Pathology of the acute phase. Br J Exp Pathol 44: 427–436

Welsh RM, Buchmeier MJ (1979) Protein analysis of defective interfering lymphocytic choriomeningitis virus and persistently infected cells. Virology 96: 503–515

Wong GHW, Bartlett PF, Clark-Lewis I, Battye R, Schrader JW (1984) Inducible expression of H-2 and Ia antigens on brain cells. Nature 310: 688–691

Strategies for Ultrastructural Visualization of Biotinated Probes Hybridized to Messenger RNA In Situ

R. H. SINGER, J. B. LAWRENCE, F. SILVA*, G. L. LANGEVIN, M. POMEROY, and S. BILLINGS-GAGLIARDI

1 Introduction

Our laboratory has emphasized in situ hybridization as a means of investigating the expression and distribution of specific mRNA molecules within single cells, particularly in developing chicken muscle cells in culture. Using an in situ hybridization protocol optimized for maintaining the structural integrity of the cell and the

Department of Cell Biology, University of Massachusetts Medical School, 55 Lake Avenue North, Worcester, MA 01605, USA.
This work was supported by NIH grant HD18066 to RHS and JBL, and by NIH grant NS11425 to SB-G and M. K. Wolf.
* Current address: Chairman, Dept. of Pathology. Univ. of Oklahoma, Health Science Center, Oklahoma City, OK 13190

native configuration of RNA (LAWRENCE and SINGER 1985; SINGER et al. 1986b), we investigated the intracellular distribution of specific mRNAs within intact myoblasts and fibroblasts and provided evidence indicating that mRNAs for certain cytoskeletal proteins are localized within specific regions of the cell (LAWRENCE and SINGER 1986). In particular, actin mRNA was found to be associated with the mobile structure of the cell, the lamellipodia, shown to be site of active actin polymerization (WANG 1985).

An understanding of the molecular mechanism(s) governing this mRNA localization will require elucidation of the physical association of mRNA with the cytoskeleton, since this may clarify how mRNAs are held in place within the cell or are transported to their final destinations. The investigation of mRNA–cytoskeleton association would be greatly facilitated by development of an electron microscopic in situ hybridization methodology. An ultrastructural study would provide a higher resolution view of the intracellular distribution of mRNA molecules particularly as it relates to the disposition of the well-studied cytoskeletal proteins.

We have extended our work on nonisotopic detection to the resolution of the electron microscope using biotin-labelled probes (SINGER et al. 1986a, 1987b, 1989; SILVA et al. 1989). High resolution is possible since the detection of biotin occurs directly at the site of hybridization. Much previous work using biotin-labelled probes detected by indirect immunocytochemistry with colloidal gold-labelled anti-bodies has been done by B. HAMKALO and her associates (e.g., HUTCHINSON et al. 1982; RADIC et al. 1987) for high resolution studies on reiterated chromosomal sequences. BINDER et al. (1986) have employed in situ hybridization using biotinated probes to lowicryl-embedded thin sections to detect mitochondrial rDNA and small nuclear RNAs using biotin antibodies and protein A–colloidal gold. WEBSTER et al. (1987) have reported the detection of P_O mRNA in Schwann cells and WOLBER et al. (1988) have reported the detection of cytomegalovirus sequences in infected fibroblasts.

Recent work (SINGER et al. 1989) has developed a method of analysis of electron microscopic data which gives extremely high signal-to-noise ratios. This approach, described in more detail later, utilizes colloidal gold antibodies to detect biotinated probe molecules hybridized to the target mRNAs. Since the mRNA acts as a template, the smaller probe molecules hybridize in an iterative way to the message strand. Analysis in this way allowed us to identify single mRNA molecules.

We undertook to continue our investigations of muscle proteins and messenger RNAs in developing myotubes, since we have analyzed the expression of similar genes in previous publications (PUDNEY and SINGER 1979, 1980; SINGER and PUDNEY 1984; LAWRENCE and SINGER 1985, 1986; SINGER et al. 1986a, b, 1987a, b), particularly in regard to the use of nonisotopic detection methods (i.e., biotinated probes). The use of biotinated oligonucleotides to investigate the expression of different isoforms of actin has revealed the asynchronous expression of particular genes representing the cardiac form of actin, the skeletal form, and the suppression of β-actin (LAWRENCE et al. 1989). Having elucidated the expression of actin, we wished to explore as well the expression of the myosin heavy-chain gene, particularly since this may shed light on the development of the sarcomere in early myogenesis. While whole critical-point dried cells viewed by electron microscopy have proven valuable for detection of cytoskeletal mRNAs in single cells in culture, the more dense structure of the myofiber necessitates a thin sectioning approach. We then extended our observations using biotin-labelled probes in conjunction with a "preembedding" ultrastructural thin sectioning approach to detect mRNA. This provided further information on message distribution and its association with certain cellular components. In particular, the investigation of the distribution of myosin heavy-chain mRNA in sectioned myotubes can elucidate the relationship of protein synthesis to the developing sarcomere. Results in an earlier publication (SILVA et al. 1989) illustrated an approach to these questions and emphasized difficulties in analyzing technical parameters based on high magnification studies.

In this chapter, we detail the strategies we have employed to approach the superimposition of precise molecular information onto high resolution ultrastructural morphology. In addition, we summarize the progression of our ultrastructural investigations from Triton-extracted single cells to Triton-extracted thin-sectioned myotubes, both of these in tissue culture; and finally we report the development of a method for in situ hybridization to fixed tissue. This method has been used to investigate the relationship of myosin heavy-chain mRNA to the developing sarcomere in chicken muscle tissue.

2 Materials and Methods

2.1 Biotination of Probes

During nick translation with biotinated nucleotide, the probe is reduced to fragments that preferably should average about 200 nucleotides per molecule. The probe size is controlled by the DNase concentration in the reaction. Probe size was monitored by alkaline agarose gel electrophoresis, using alkaline phosphatase detection on probes transferred to nitrocellulose (BRL, Gaithersburg, MD, DNA detection kit). Incorporated into this probe was a biotinated analog of thymidine [biotin-11-deoxyuridine triphosphate (dUTP) Enzo Biochemicals New York, N.Y] with several biotins per 100 nucleotides. Biotin-specific activity was measured by incorporating phosphorus-labelled deoxycytidine triphosphate (dCTP) into the probe simultaneously with biotinated nucleotide [60 µm, diluted 1:10 with unlabelled thymidine triphosphate (TTP)]. Assuming the biotinated nucleotide was incorporated at 80% of the rate of the equivalent unlabelled nucleotide (LANGER et al. 1981) and that the total incorporation of dCTP was equimolar to TTP, we calculate that the nick-translated probe was substituted approximately 2% with biotinated nucleotides.

Probes used were the full-length β-actin clone (1.9 kb) and the full-length sarcomeric myosin heavy-chain clone (a gift of J. Robbins, 5.9 kb).

2.2 Treatment Through Hybridization

2.2.1 Cells for Whole-Mount Electron Microscopy

2.2.1.1 Cell Culture

Skeletal myoblasts and fibroblasts were isolated from the pectoral muscles of 12-day-old chicken embryos and cultured by standard techniques. Cells were plated at a density of 2×10^5/well in Costar six-well tissue culture plates containing grid assemblies: carbon–formvar-coated electron microscope grids on 22-mm circular coverslips. Grids were London 200 mesh or honeycomb gold finder types (Fullam, Laham, N.Y.). Carbon rods and formvar were purchased from Tousimis (Rockville, MD). Grid assemblies were made by floating a formvar film onto glass coverslips supporting between four and nine gold grids. After drying, the formvar ist stabilized with a thin layer of carbon. Grid assemblies in multiwell dishes were sterilized by γ irradiation from a cesium 137 source in a Gammcell (Atomic Energy of Canada, Ltd.). All pretreatments and washes were performed in the six-well tissue culture plates.

Three-day old cultures were extracted and fixed by a modification of the method of LENK et al. (1977). Cells were briefly washed with isotonic buffer (0.3 M sucrose, 0.1 M NaCl, 10 mM 1,4 Piperazinediethanol sulfonic acid (PIPES), 3 mM MgCl$_2$, 10 µM Leupeptin (Sigma, St. Louis, MO). 1:40 dilution of vanadyl complex (4 mM adenosine, 0.2 M VOSO$_4$, pH 7), then extracted for 90 sec in the same buffer plus 0.5% Triton X-100 (Boehringer Mannheim, W. Germany) followed by another brief wash in the first buffer, fixed with 4% glutaraldehyde in phosphate-buffered saline, 5 mM magnesium, and passed through an ethanol series (30%, 50%, 70% for 5 min each). Each solution was removed by aspiration. Cells were stored in 70% ethanol at 4 °C until used. In order to preserve microtubes, all solutions were used at room temperature and ethylene glycol tetra-acetic acid was included to eliminate calcium.

2.2.1.2 Hybridization

The details and derivation of the hybridization protocol have been published elsewhere (LAWRENCE and SINGER 1985; SINGER et al. 1986b, 1987a). The salient features of this method are that cell treatments that disrupt morphology or cause loss of mRNA have been minimized. However, Triton extraction of the cells was used initially in order to facilitate satisfactory probe penetration (see Sect. 3). Briefly, cells on grid assemblies were rehydrated in phosphate saline with 5 mM MgCl$_2$ for at least 10 min, followed by 0.1 M glycine, 0.2 M TRIS, pH 7.4 for 10 min. Afterwards cells were kept in 50% formamide, 2X SSC at 37 °C for 15 min prior to hybridization. For ribonuclease controls, 100 µl of a 100 µg/ml solution of RNase A in 2X SSC was applied to samples for 1 h at 37 °C prior to the 50% formamide 2X SSC pre-hybridization treatment. Probes used were biotinated by nick translation (LANGER et al. 1981) and were approximately 100–200 nucleotides in length. Cells were hybridized in buffer containing 50% formamide, 2X SSC with 100 µg/µl each of tRNA and sheared salmon sperm DNA, 1% bovine serum albumin (BSA) and 1 µg/ml of probe. The nucleic acid components were heated to 90 °C for 5 min in 100% formamide before adding the other components.

Each grid assembly was placed cell-side down onto a 10 µl drop of the hybridization mixture on parafilm, covered with an additional sheet of parafilm to prevent drying, and incubated for 3 h in a humidified chamber of 37 °C. After hybridization, coverslips were gently removed from the parafilm and placed in individual wells in a clean multiwell dish containing 2.5 ml 50% formamide, 1X SSC, and finally 1X SSC, each for 30 min. The cells were left in 1X SSC overnight at 4 °C to be stained the next day with antibodies.

2.2.2 Cultured Myotubes for Thin-Section Electron Microscopy

2.2.2.1 Cell Culture

Cells grown on glass or plastic coverslips were used, Triton-extracted as described above, and postfixed in 2% glutaraldehyde; other fixatives, when used appropriately, can also give good results (SILVA et al. 1989). The cells were either processed immediately after fixation or stored in 70% ethanol or phosphate-buffered saline for brief periods of time (days) if antibodies to protein were to be used.

2.2.2.2 Hybridization

Cells were rehydrated in phosphate-buffered saline plus 5 mM MgCl$_2$ for 10 min in Coplin jars, followed by 0.1 M glycine, 0.2 M TRIS-HCl, pH 7.4, for 10 min. The cells were then placed in 50% formamide (Fluka, Switzerland) 2X SSC (0.3 M NaCl in sodium citrate buffer) for 10 min prior to hybridization. The probe, $E.$ $coli$ tRNA, and salmon sperm DNA were lyophilized and then resuspended in formamide and melted at 90 °C for 10 min. Just prior to placing on the cells, the probe, tRNA, and DNA were combined with the hybridization mix, so that the final probe concentration was 1 µg/ml and the final hybridization solution consisted of 50% formamide 2X SSC, 0.2% BSA, 10 mM vanadyl sulfate ribonucleoside complex (BERGER and BIRKENMEIER 1979; this can be eliminated if all solutions are autoclaved and RNase-free), 10% dextran sulfate (Sigma), and 1 mg/ml each of $E.$ $coli$ tRNA and salmon sperm DNA. Cells on the coverslips were incubated in 20 µl hybridization solution for 3 h at 37 °C by putting the coverslips cell-side down on parafilm. After hybridization, coverslips were placed in 10-ml Coplin jars (VWR) and rinsed three times with shaking for 30 min each in 2X SSC, 50% formamide at 37 °C; SSC, 50% formamide at 37 °C; and SSC at room temperature.

2.2.3 Aldehyde-Fixed Developing Muscle for Thin-Section Electron Microscopy

2.2.3.1 Tissue Preparation

Hand-cut thin strips of pectoral muscle from 14-day-old chicken embryos were immersed in 2% paraformaldehyde 0.2% glutaraldehyde for 15 min. They were then rinsed and extracted for 15 min in 0.5% Triton and 0.5% saponin, followed by 0.1 M glycine, 0.2 M TRIS-HCl (pH 7.4) for 10 min. Strips were placed in 50% formamide (Fluka), in 0.3 M NaCl in sodium citrate buffer for 10 min.

2.2.3.2 Hybridization

Strips were incubated overnight in individual 20 µl drops of a hybridzation solution which differed from that used for cultured cells only by the absence of VOSO$_4$ ribonucleoside complex.

2.3 Detection of Hybridization

2.3.1 Cells on Grids or Coverslips

Grids or coverslips were pretreated with 8% RNase-free BSA (Boehringer, FRG), TRIS-buffered saline (TBS), pH 7.4, for 10 min. Excess fluid was removed with bibulous paper and each grid was transferred to 100 µl drops of the primary antibody staining solution: rabbit antibiotin (Enzo Biochemicals) diluted 1:100 in 1% BSA, 0.5% Triton X-100, TRIS-buffered saline, pH 7.4 (BTTBS) and incubated for 2.5 h at room temperature. They were washed for 30 min on a stirring platform with three changes of 3 ml/grid or coverslip of the same buffer. Incubation with the secondary antibody, colloidal gold-conjugated goat antirabbit (GAR 10, Janssen Pharmaceutica, Belgium) diluted 1:2 to 1:10 in the same BTTBS buffer, was overnight at room temperature. The secondary antibody was centrifuged on a tabletop centrifuge for 9 min (tap 2) to remove gold clusters before using. When protein was labelled as well as the mRNA, mouse monoclonal antibodies were used in conjugation with a 5-nm colloidal gold-conjugated goat-antimouse antibody (Janssen). For double labelling procedures, both primary (antiprotein and antibiotin) and both secondary antibodies (antimouse and antirabbit) were treated simultaneously. The grid assemblies or coverslips were placed in multiwell dishes on parafilm and sealed tightly to prevent evaporation. After incubation, the coverslips were washed in PBS three times at room temperature. They were then placed in 2% glutaraldehyde in PBS for 2 h at room temperature and then rinsed in PBS for 10 min (three times) at room temperature on a stirring platform in the multiwell dishes.

2.3.2 Muscle Tissue

Detection methods were identical to those described for cells, except that primary and secondary antibodies were used undiluted and the time in primary antibody was increased to 4 h.

2.4 Electron Microscopy

2.4.1 Whole-Mount Cells

Cells on the grid assemblies were critical-point dried after dehydration in graded ethanol and observed with a JEOL 100S or 100CX electron microscope at an accelerating voltage of 100 kV without further contrast enhancement.

2.4.2 Cells for Thin Sectioning

Coverslips containing hybridized myotubes were postfixed in freshly dissolved 2% OsO4 in Sørensen's phosphate buffer for 30 min. After rinsing in distilled H_2O (four times, 10 min each, room temperature), they were dehydrated in a series of graded ethanol (50%, 70%, 95%, twice each for 5 min). They were then placed in fresh 100% alcohol three times for 10 min each. After treatment with 100% alcohol/fresh Epon (1:1) for 30–60 min at room temperature, they were placed in 100% Epon for 1 h at room temperature. Beem capsules were filled with fresh 100% Epon and quickly inverted onto the surface of the coverslips and oven baked for 48 h at 60 °C. The petri dish was immersed in liquid nitrogen for a few seconds, then gently but firmly tapped from behind to remove the Beem capsule and attached coverslip (this was much easier with glass than with plastic coverslips). The Beem capsules were then routinely processed for ultrathin sectioning. Transmission electron microscopy was then performed on a JOEL 100S electron microscope at an accelerating voltage of 60 kV.

2.4.3 Muscle Tissue

Strips were further fixed after hybridization in 2% glutaraldehyde for 1 h, rinsed briefly in buffer, dehydrated in a graded series of ethanols, infiltrated, and embedded in Epon. The tissue was initially cut in strips rather than minced to aid in orientation when embedding and sectioning.

2.5 Quantitative Electron Microscopic Studies on Cultured Cells

A standardized approach was instituted for the quantitation of the 10-nm gold particles counted from the electron microscopic prints. Multiple cells on grids or multiple sections placed on grids were studied from each experiment. Single cells in whole-mount or well-developed myotubes (those with obvious myofiber development) were studied and micrographs at a standard magnification were taken from regions showing the greatest gold concentrations. The gold particles were manually counted from each print and the total number of clusters of 2, 3, 4 particles etc., was also determined. Gold particles at the edge of the cell (i.e., at the cell membrane) were not counted. Quantitation of gold particles was expressed in mean ± standard deviation and statistical analysis was performed utilizing the Student t test determination. In addition, values are expressed using the total number of gold particles counted as clusters. A large number of prints were studied for quantitation in each experiment.

3 Results

3.1 In Situ Hybridization Visualized in Triton-Extracted Whole Mounts of Single Cells

Isotopic experiments done previously (LAWRENCE and SINGER 1985) indicated a signal-to-noise ratio greater than 50:1 for hybridization to actin mRNA even in Triton-extracted cells (SINGER et al. 1989). This demonstrated that the extraction, fixation and hybridization parameters used provided a strong hybridization signal, appropriate for pursuing further experiments using nonisotopic detection of biotin-labelled probes. The use of nonisotopic detection would be expected to reduce the signal-to-noise ratio, since each additional step has the potential to introduce background noise and to diminish the efficiency of detection. Since the gold particles give a pictorial representation of in situ hybridization (such as silver grains do in autoradiography), and are quantifiable, they allow a more rigorous analysis of signal to noise. The method of data analysis was established in initial experiments, all of which utilized Triton-extracted whole mounts of chicken fibroblasts and myoblasts hybridized with an actin probe. The most direct method is to count gold particles within cells on experimental (biotinated actin probe) samples compared to those within cells on negative control samples (biotinated pBR322 probe without actin insert, hybridized under identical conditions). Quantitation of overall gold densities for at least 12 cells in four experiments demonstrated an average signal-to-noise ratio higher than 6:1. Therefore, when the actin probe is used, more than 80% of the total gold particles represent bona fide hybridization. Further inspection of gold particle distribution and theoretical considerations of labelling and detection made it evident that individual sites of hybridization could be identified with very high confidence. Since our intent is to visualize hybridization with nanometer resolution at the high magnification of the electron microscope, this is a key element in the methodology and subsequent data analysis. The strategy derives from the fact that many of these small probe molecules hybridize along the larger mRNA template, resulting in a tightly clustered array of gold particles qualitatively distinct from adventitious noise.

Since the target site for in situ hybridization was the β-actin mRNA of 1816 nucleotides (KOST et al. 1983), or the myosin heavy chain; 6 kb (MOLINA et al. 1987; see Sects. 3.2, 3.3), many probe molecules will hybridize at random positions, *dependent* on the target mRNA. Subsequent detection of biotin by indirect immunocytochemistry would result in this iterated array of biotins being detected by a parallel array of antibodies, thereby producing consecutive electron-dense 10-nm images representing the templating of the target molecule. Noise, however, would have different characteristics, because any nonspecific sticking of the small probe molecules or of the antibodies would be very unlikely to form an array of gold particles, since they would be individual events *independent* of the templating effect of the target molecules. Hence, the signal would be expected to be both quantitatively and qualitatively different from noise. Statistical evidence presented elsewhere (SINGER et al. 1989) supports this interpretation.

Antibody background noise was measured by following the protocol but without the probe. The number of colloidal gold particles per unit cellular area in control samples reacted with biotinated pBR322 probe was significantly less than in samples reacted with pBR322 containing either myosin or actin insert. More importantly, the presence of hybridizing probe increased significantly the number and size of the colloidal gold *clusters*. The largest number of gold particles were single, and for single particles signal-to-noise ratios were seen to be about 3:1. However, for increased number of particles in a cluster, progressively fewer clusters were observed using the control probe relative to the experimental probe. It was apparent that hybridization with the probe sequence yields a significant number of larger-sized (greater than 6) clusters, such that sizable clusters of gold-labelled antibodies are detected. This results in the expected proportional increase in signal-to-noise ratio increasing logarithmically to 30:1 for clusters of eight colloidal gold particles. Therefore, when as many as eight particles are seen in a cluster, the probability is 97% that this signal results from the dependent interaction of the probe and the mRNA.

When the individual clusters were examined, it became evident that there were additional qualitative differences between the mRNA probe and the control probe. Significant numbers of clusters were found where the gold particles were not simply in an aggregate, but rather in a characteristic array with relatively constant spacing and a linear dimension proportional to the number of gold particles involved. This is consistent with the theory of the probe hybridizing to a template. Examples of the visualization of individual clusters denoting hybrids are given in Fig. 1. This micrograph shows the arrangement of gold particles in clusters for which the number of particles is such that it represents bona fide hybridization to an mRNA molecule. The structure of the cluster appears to be circular. In many cases, circular arrangements prove to be actually spiral when examined by stereo microscopy

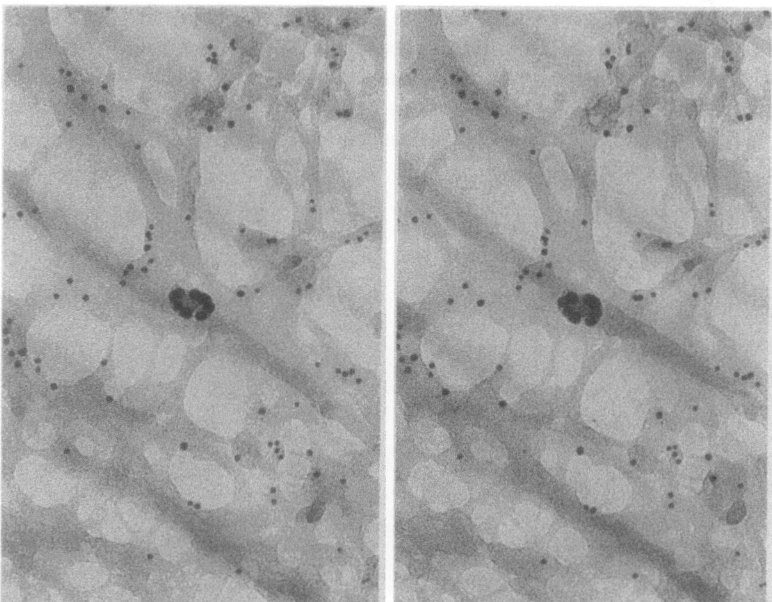

Fig. 1. Stereo pair of β-actin mRNA detected by in situ hybridization. Actin protein was detected in the same samples using 5-nm gold conjugated antibodies indirectly. Triton-extracted myoblasts fixed in glutaraldehyde were used for the hybridization, after having been grown on grids. Biotinylated actin DNA probe was used and detected by 10-nm colloidal gold conjugated antibodies using indirect immunocytochemistry. × 165 000

(a stereo pair is provided for illustration). The sample has also been exposed to monoclonal antibodies to actin which have been detected by indirect immunocytochemistry using 5-nm colloidal gold. This allows the high-resolution visualization of both the message and its cognate protein in the same area. It is evident that actin protein is distributed throughout the area occupied by the mRNA. The association of the messages with the cytoskeleton can be seen clearly in these preparations. In almost all cases examined, the mRNA detected sits astride several filaments, very often at the base of, or along, a long, single filament with the diameter of actin filaments (6 nm).

3.2 In Situ Hybrids Visualized in Triton-Extracted, Cultured Myotubes: A Preembedment Procedure

We found that the Triton extraction followed by glutaraldehyde fixation afforded a satisfactory combination of probe penetration and preservation of the structure of our particular interest: the developing sarcomere. It should be emphasized that membranes or membrane-associated structures are solubilized in this procedure. The particular method emphasizes a preembedment technique. It follows from previous work on whole-mount cells that, in order to investigate the developing myotube — a structure too dense to penetrate in its entirety with the electron beam —

it would be necessary to section this structure after hybridization. In contrast to previous work, we used a probe for myosin heavy chain instead of actin, because the 6 kb message offered a bigger target and thus would be more likely to withstand the signal reduction involved in partitioning the myotube into thin sections. In addition, whether myosin mRNA would be associated with developing sarcomeres is, in itself, an interesting question.

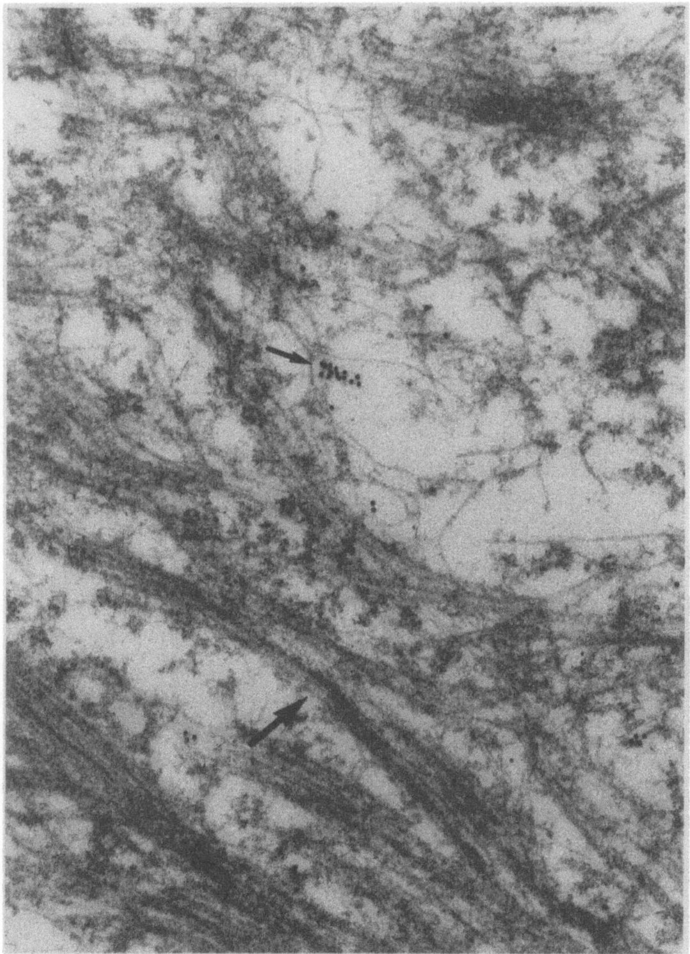

Fig. 2. Detection of myosin heavy-chain mRNAs in Triton-extracted myotubes, hybridized and thin sectioned. Cultured myotubes were Triton-extracted and fixed with 2% glutaraldehyde for 15 min at room temperature before hybridization and sectioning. A cluster of colloidal gold particles can be seen associated with an electron-dense filamentous protein of the diameter of myosin (*small arrow*). The number of colloidal gold particles indicates that it is a myosin heavy-chain message (see text), presumably in the act of translating a myosin protein. A developing myofiber is within 260 nm (*large arrows*). ×77000

The quantitative approach to the data analysis was identical to that described previously which involved the counting of colloidal gold particles used to detect the biotinated probe, both individually and as various cluster sizes. This allowed a critical evaluation of signal-to-noise ratio when comparing the MHC probe cloned into pBR322 versus the vector alone.

Besides the numbers of colloidal gold particles, morphology of the cells was assessed in each sample, as well as the localization of the signal, i.e., if myosin mRNA was located near assembling myofibers. Finally, the size of the colloidal gold clusters was evaluated for evidence of templating onto messages.

Cluster conformations appear to be consistent with the idea of hybridization to a message template. Increase in cluster size with the MHC probe was significantly greater than with the control probe in most experiments. Although this method requires extensive analysis of sections, it yields images which give a high degree of certainty that a message for myosin is visualized. In many of these images, the message is associated with an electron-dense filamentous structure which may have the characteristics of myosin (i.e., diameter of 14 nm). A good double-labelling protocol would be more conclusive in identifying this filament. Because of the sectioning of the tissue, it is difficult to assess the three-dimensional structure of the message, as was done on whole-mount cells.

Figure 2 illustrates an example of the results obtained from this approach. Sarcomeres in developing myofibrils can be easily seen in the region of the mRNA detected, but the area immediately surrounding consists of loose filamentous material where the signal for myosin mRNA is predominant. The mRNA can be seen associated with what appears to be a growing myosin polymer.

Recent work by ISAACS and FULTON (1987) and FISCHMAN (personal communication) have shown that a population of myosin heavy chains is either cotranslationally assembled into filaments, or is very rapidly assembled after translation. We have observed evidence of clusters of gold (i.e., mRNA for MHC) near or on filaments which may be in the process of assembly. Since these represent very few examples, most likely the result of fortuitous sections, it is not possible to ascertain whether or not this is a general occurrence. The approach, however, illustrates the potential for characterizing the relationship between the synthesis of cellular filament systems and their assembly, such as occurs during sarcomerogenesis.

3.3 In Situ Hybridization on Unextracted Tissue: Preembedment Procedure

Work currently underway in our laboratories involves the extension of this approach to muscle tissue from developing embryos. Enhancement of signal-to-noise ratios has been a major concern, since quantitation of many samples is extremely time-consuming and cumbersome. Obviously, if the control probe gave no background at all, signal would be significant at any level and many fewer samples would require screening!

Preliminary quantitation suggests a signal-to-noise ratio of at least 6:1 using the methods described here for fixed tissue. Two advances are made by this procedure: the first is that fixation prior to detergent extraction produces morphological preservation more comparable with conventional methods, and circumvents the

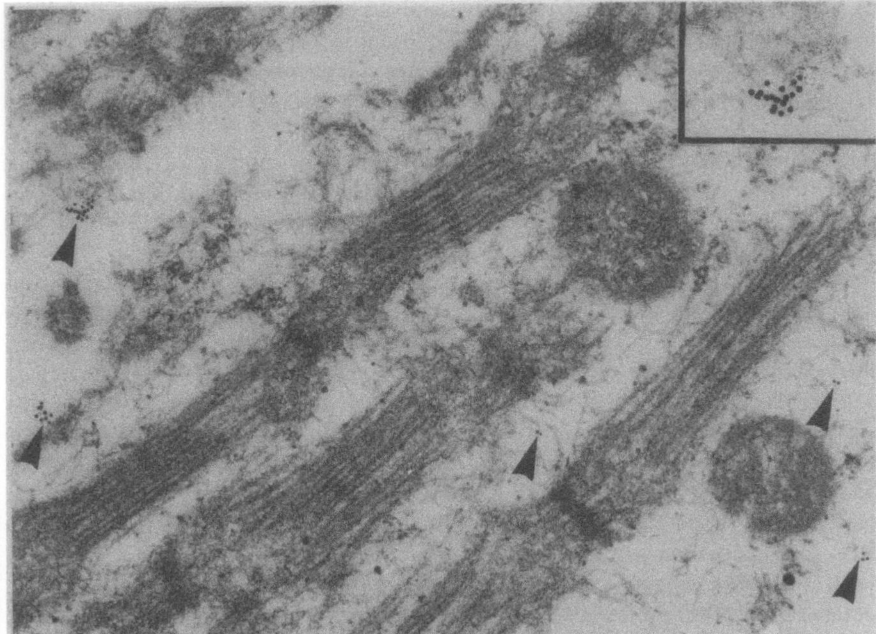

Fig. 3. Detection of myosin heavy-chain mRNA in fixed tissue. Pectoral muscle tissue from 14-day-old chicken embryos was fixed prior to detergent extraction and hybridization with myosin heavy-chain probe. Colloidal gold particles, which almost certainly indicate the position of myosin heavy-chain message, are associated with cytoplasmic filaments adjacent to developing myofibrils (*arrow heads*). At higher magnification, a group of 18 gold particles in this section has the configuration of a twisted string. Controls probe showed no clusters of a size greater than three colloidal gold particles. × 36000; inset × 72000

potential problem of structural rearrangement during detergent treatment of unfixed cells. The second is that molecular information is obtained from cells developing in vivo rather than in vitro. Therefore, possible artifacts of tissue culture are eliminated while maintaining the quality of the molecular results. Figure 3 illustrates the quality of the data obtainable for tissue material prepared and hybridized as in Sect. 2. Much the same results are obtained as with Triton-extracted myotubes, i.e., the mRNA appears to be intimately associated with cytoplasmic filaments immediately adjacent to developing myofibrils. However, in this case, the events described are occurring in developing chicken muscle in ovo and the morphological context of these events is much better preserved.

4 Summary

A progressive development of the application of in situ methodology to ultrastructural procedures has resulted in the ability to detect individual molecules of mRNA with high probability. Beginning with whole-mount cells and then developing myotubes, both in culture and detergent extracted before fixation, we were able to progress to methods which allow detection of mRNA in tissue sections. Initial results confirm that the detection of mRNA in thin-sectioned tissue is very similar to observations on the extracted, cultured cells, and that the same methods of data analysis apply. Current work is devoted to the application of the methodology to other cellular structures, such as the nucleus, and to other tissue-probe systems, such as brain.

Acknowledgements. The authors appreciate the skilled help from John McNeil and Shirwin Pockwinse in the laborious and time-consuming preparations of material and photography. FS was on sabbatical leave from the Department of Pathology at Southwestern Medical Center.

References

Ben-Ze'ev A, Horowitz M, Skolnik H, Abulafia R, Laub O, Aloni Y (1981) The metabolism of SV40 RNA is associated with cytoskeletal framework. Virology 111: 475–487

Berger SL, Birkenmeier CS (1979) Inhibition of intractable nucleases with ribonucleoside-vanadyl complexes: isolation of mRNA from resting lymphocytes. Biochemistry 18: 5139–5143

Binder M, Tourmente S, Roth J, Renaud M, Gehring W (1986) In situ hybridization at the electron microscopic level: Localization of transcripts on ultrathin sections of lowicryl K4M-embedded tissue using biotinated probes and protein A-gold complexes. J Cell Biol 102: 1646–1653

Bonneau AM, Darveau A, Sonnenberg N (1985) Effect of viral infection on host protein synthesis and mRNA association with the cytoplasmic cytoskeletal structure. J Cell Biol 100: 1209–1218

Cervera M, Dreyfuss G, Penman S (1981) Messenger RNA is translated when associated with cytockeletal framework in normal and VSV-infected HeLa cells. Cell 23: 113–120

Farmer SR, Ben-Ze'ev A, Benecke BJ, Penman S (1978) Altered translatability of messenger RNA from suspended anchorage-dependent fibroblasts: reversal on cell attachment to a surface. Cell 15: 627–637

Fulton AB, Wan KW, Penman S (1980) The spatial distribution of polyribosomes in 3T3 cells and the associated assembly of proteins into the skeletal framework. Cell 20: 849–857

Howe JG, Hershey JWB (1984) Translational initiation factor and ribosome association with the cytoskeletal framework fraction from HeLa cells. Cell 37: 85–93

Hutchison NJ, Langer-Safer PR, Ward DC, Hamkalo BA (1982) In situ hybridization at the electron microscope level: hybrid detection by autoradiography and colloidal gold. J Cell Biol 95: 609–618

Isaacs WB, Fulton AB (1987) Cotranslational assembly of myosin heavy chain in developing cultured skeletal muscle. Proc Natl Acad Sci USA 84: 6174–6178

Jamrich MKA, Mahon KA, Gavis ER, Gall JG (1984) Histone RNA in amphibian oocytes visualized by in situ hybridization to methacrylate-embedded tissue sections. EMBO J 3: 1939–1943

Jeffrey WR (1984) Spatial distribution of messenger RNA in the cytoskeletal framework of Ascidian eggs. Dev Biol 103: 482–492

Kost TA, Theodorakis N, Hughes SH (1983) The nucleotide sequence of the chicken cytoplasmic beta actin. Nucleic Acids Res 11: 8287–8301

Langer RR, Waldrop AA, Ward DC (1981) Enzymatic synthesis of biotin-labeled polynucleotides: novel nucleic acid affinity probes. Proc Natl Acad Sci USA 78: 6633–6637

Lawrence JB, Singer RH (1985) Quantitative analysis of in situ hybridization methods for the detection of actin gene expression. Nucleic Acids Res 13: 1777–1799

Lawrence JB, Singer RH (1986) Intracellular localization of messenger RNA for cytoskeletal proteins. Cell 45: 407–415

Lawrence JB, Taneja K, Singer RH (1989) Temporal resolution and sequential expression of muscle-specific genes revealed by in situ hybridization Dev Biol 132

Lenk R, Penman S (1979) The cytoskeletal framework and poliovirus metabolism. Cell 16: 289—301

Lenk R, Ransom L, Kaufmann Y, Penman S (1977) A cytoskeletal structure with associated polyribosomes obtained from Hela cells. Cell 10: 67–78

Manuelidis L (1985) Indications of centromere movement during interphase and differentiation. Ann NY Acad Sci 450: 205–221

Manuelidis L (1985b) In situ detection of DNA sequences using biotinylated probes. Focus 7: 4–8

Manuelidis L, Ward DC (1984) Chromosomal and nuclear distribution of the *Hind*III 1.9 kb human DNA repeat segment. Chromosoma 91: 28–38

Manuelidis L, Langer-Safer PR, Ward DC (1982) High resolution mapping of satellite DNA using biotin-labeled DNA probes. J Cell Biol 95: 619–625

McBeath E, Fujiwara K (1984) Improved fixation for immunofluorescence microscopy using light-activated 1,3,5-triazido-2,4,6-trinitrobenzene (TTB). J Cell Biol 99: 2061–2073

Molina MI, Kropp KE, Gulick J, Robbins J (1987) The sequence of an embryonic myosin heavy chain gene and isolation of its corresponding cDNA. J Biol Chem 262: 6478–6488

Pudney J, Singer RH (1979) Electron microscopic visualization of the filamentous reticulum in whole cultured presumptive chick myoblasts. Am J Anat 156: 321–336

Pudney J, Singer RH (1980) Intracellular filament bundles in whole mounts of chick and human myoblasts extracted with Triton X-100. Tissue Cell 12: 595–612

Radic MZ, Lundgren G, Hamkalo BA (1987) Curvature of mouse satellite DNA and condensation of heterochromatin. Cell 50: 1101–1108

Silva FG, Lawrence JB, Singer RH (1989) Progress toward ultrastructural identification of individual mRNAs in this section: myosin heavy chain in developing myotubes. In: Bullock, G and Petrusz P (ed) Techniques in immunocytochemistry. Vol. 4, Academic Press, London

Singer RH, Pudney J (1984) Filament-directed intercellular contacts during differentiation of cultured chick myoblasts. Tissue Cell 16: 17–29

Singer RH, Langevin GL, Lawrence JB (1986a) Electron microscopic visualization of single intracellular actin messenger RNA molecules by in situ hybridization. J Cell Biol 103: 315a

Singer RH, Lawrence JB, Villnave C (1986b) Optimization of in situ hybridization using isotopic and non-isotopic detection methods. Biotechniques 4: 230–250

Singer RH, Lawrence JB, Rashtchian RN (1987a) Toward a rapid and sensitive in situ hybridization methodology using isotopic and non-isotopic probes. In: Valentino K, Eberwine J, Barchas J (eds) In situ hybridization: application to the central nervous system. Oxford University Press, New York, pp 71–96

Singer RH, Lawrence JB, Langevin GL, Rashtchian RN, Villnave CA, Cremer T, Tesin D, Manuelidis L, Ward DC (1987b) Double labelling in situ hybridization using non-isotopic and isotopic detection. Acta Histochem Cytochem 20: 589–598

Singer RH, Langevin GL, Lawrence JB (1989) Ultrastructural visualization of cytoskeletal mRNAs and their associated proteins using double label in situ hybridization (in press)

van Venrooij JJ, Sillekens PTG, van Eekelen CAG, Reinders RT (1981) On the association of mRNA with the cytoskeleton in uninfected and adenovirus-infected human KB cells. Exp Cell Res 135: 79–91

Wang YL (1985) Exchange of actin filaments at the leading edge of living fibroblasts: possible role of tread milling. J Cell Biol 101: 597–602

Webster H de F, Lamperth L, Favilla JT, Lemke G, Tesin D, Manuelidis L (1987) Use of a biotinylated probe and in situ hybridization for light and electron microscopic localization of Po mRNA in myelin-forming Schwann cells. Histochemistry 86: 441–444

Wolber RA, Beals TF, Lloyd RV, Maassab HF (1988) Ultrastructural localization of viral nucleic acids by in situ hybridization. Lab Invest 59: 144–151

Haase A, Brahic M, Stowring L, Blum H (1984) Detection of viral nucleic acids by in situ hybridization. Methods Virol 7: 189–226

Javier RT, Stevens JG, Dissette VB, Wagner EK (1988) A Herpes simplex virus transcript abundant in latently infected neurons is dispensable for establishment of the latent state. Virology 166: 254–257

Knotts FB, Cook ML, Stevens JG (1973) Latent herpes simplex virus in the central nervous system of mice and rabbits. J Exp Med 138: 740–744

McClennon JL, Darby G (1980) Herpes simplex virus latency: the cellular location of virus in dorsal root ganglia and the fate of the infected cell following virus activation. J Gen Virol 51: 233–243

Rock DL, Nesburn AB, Ghiasi H, Ong J, Lewis TL, Lokensgard JR, Wechsler SL (1987) Detection of latency-related viral RNAs in trigeminal ganglia of rabbits latently infected with herpes simplex virus type 1. J Virol 61: 3028–3826

Spivak JG, Fraser NW (1987) Detection on herpes simplex virus type I transcripts during latent infection in mice. J Virol 61: 3841–3847

Stevens JG (1975) Latent characteristics of selected herpesviruses. Adv Cancer Res 26: 227–256

Stevens JG, Cook ML (1971) Latent herpes simplex virus in spinal ganglia of mice. Science 173: 843–845

Appendix

Figure on page 12

Fig. 1a–d. Determining TMEV host range in CNS with combined immunocytochemistry-in situ ▶
hybridization. **a** Double immunostaining with anti-GFAP (alkaline phosphatase, *blue*) and anti-
CA II (horse radish peroxidase, *brown*) antisera. **b** Combined immunoperoxidase in situ hybridization.
The serum used was anti-GFAP. The *arrowheads* point to GFAP positive astrocytes and the
arrow to an infected cell with TMEV RNA **c. d** Combined immunoperoxidase in situ hybridization.
The serum used was anti-CA II. *Arrowheads* point to CA II positive oligodendrocytes. The arrow points
to infected oligodendrocytes. *Dots* are adjacent to inflammatory infiltrates characteristic of demyelinating lesions. (From AUBERT et al. 1987, with permission)

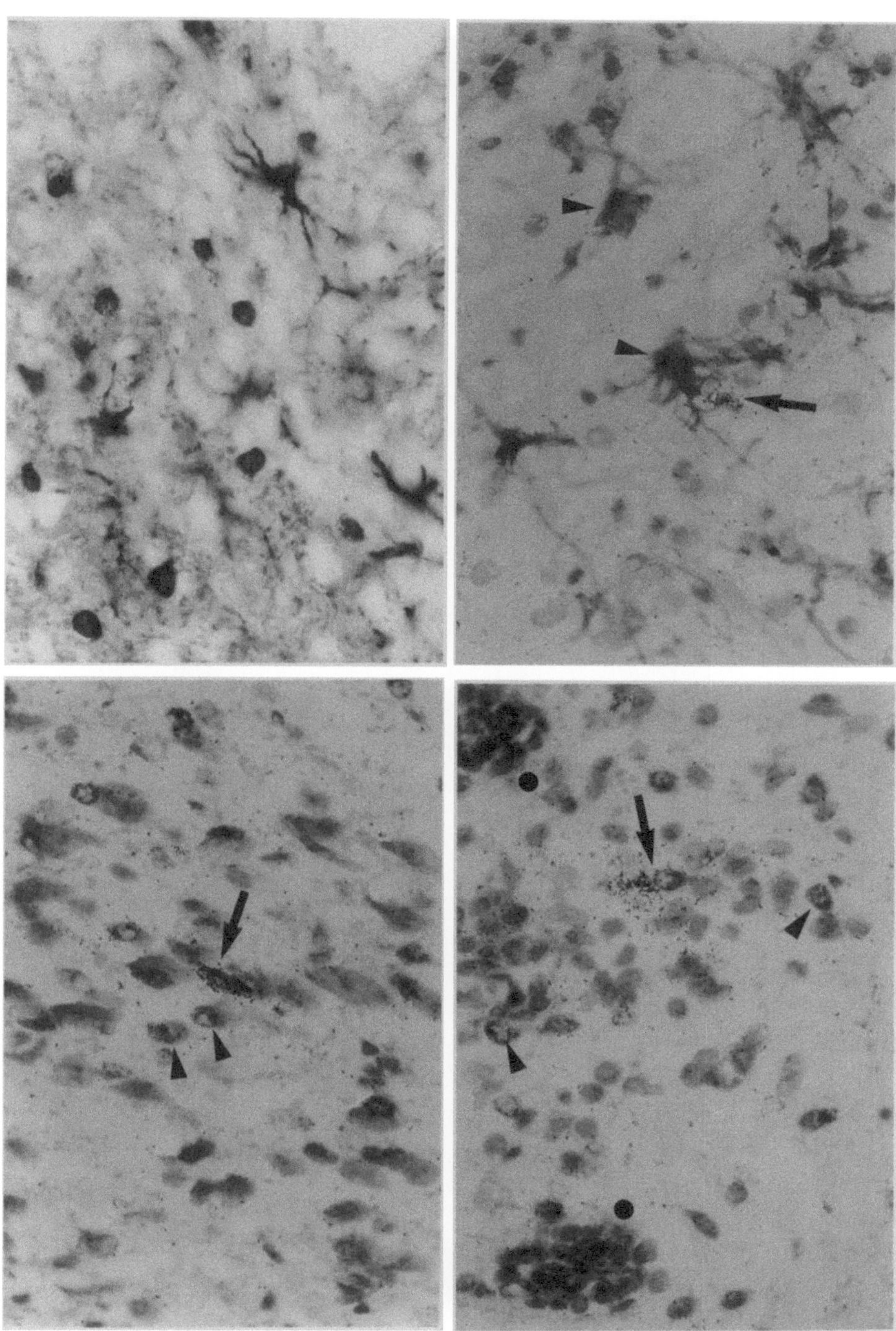

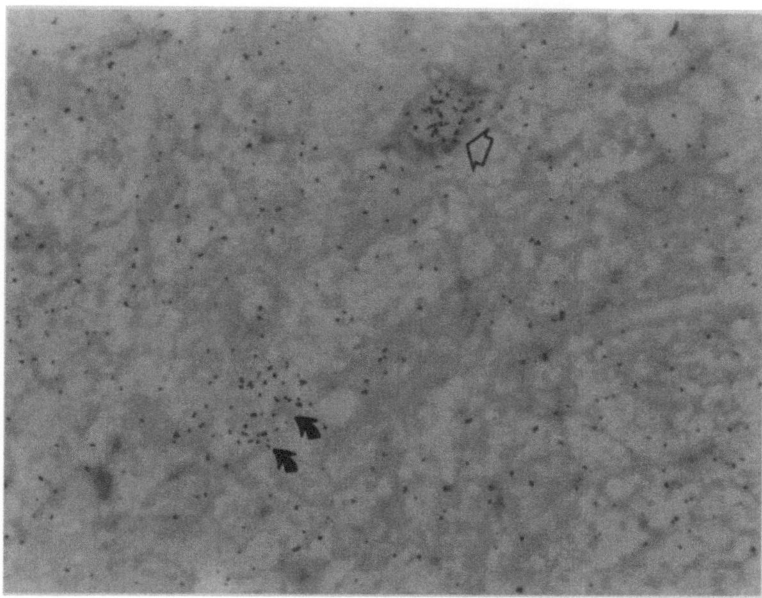

Fig. 2. Visna virus RNA in oligodendrocytes by immunocytochemistry and in situ hybridization. Some of the oligodendrocytes (*open arrow*) marked by the brown diaminobenzidine reaction product contain the virus RNA (*silver grains*), while others do not (*filled arrows*). (From STOWRING et al. 1985, with permission)

Figure on page 14

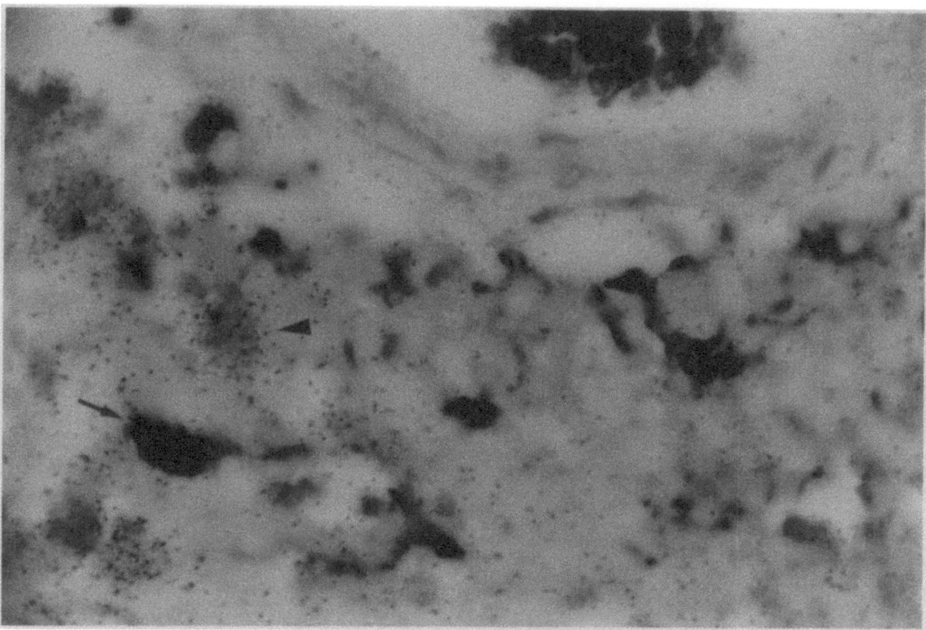

Fig. 3. HIV RNA in cells in the nervous system. Formalin-fixed and paraffin-embedded sections from an individual with AIDS dementia complex were hybridized to a HIV-specific DNA probe radiolabeled by nick translation with ^{35}S precursors. Following hybridization the sections were reacted with antibodies to the astrocyte-specific protein GFAP, biotinylated antispecies antibody, ABC peroxidase, PPD and H_2O_2. Cells of the monocyte lineage (*arrowhead*) in the inflammatory infiltrate bordering the blood vessel at the top of the figure contain HIV RNA whereas the astrocytes (*arrow*) do not. Sample viewed in transillumination and incident polarized light to reveal grains against darker staining background

Figure on page 15

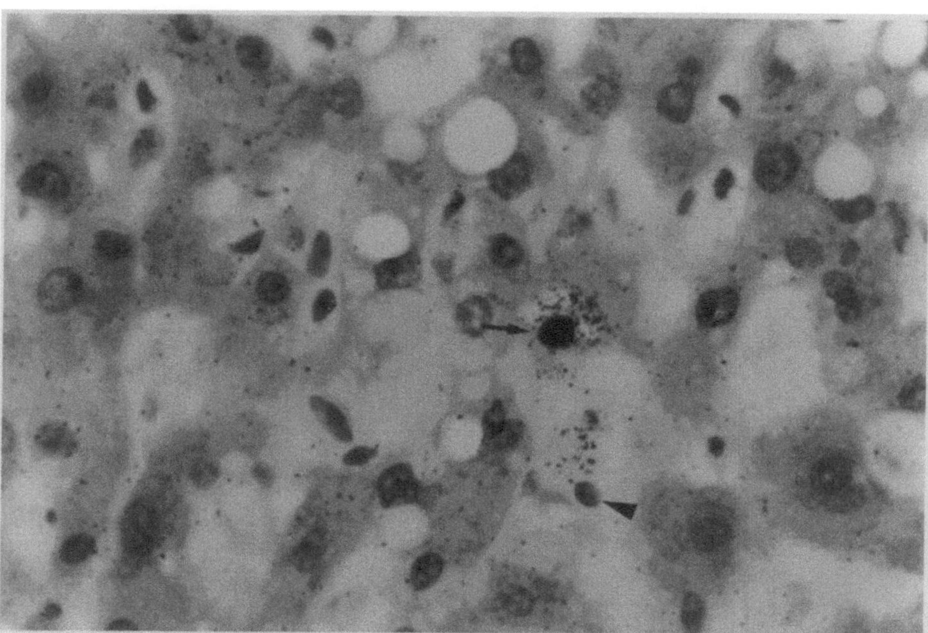

Fig. 5. Simultaneous detection of HBV core antigen and DNA in persistent infections. Formalin-fixed and paraffin-embedded sections of virus from an individual with chronic active hepatitis virus infection were reacted with antibody to HBV core antigen followed by immunocytochemical development using PPD as a chromogen. After the pretreatments described in the text to enhance hybridization, the sections were hybridized to a ^{3}H-labeled HBV-specific probe. The *arrow* points to a cell with HBV core antigens in the nucleus and viral DNA in the cytoplasm. An adjacent cell (*arrowhead*) with much less DNA is not stained for core antigen. (From BLUM et al. 1984, with permission.)

Figure on page 17

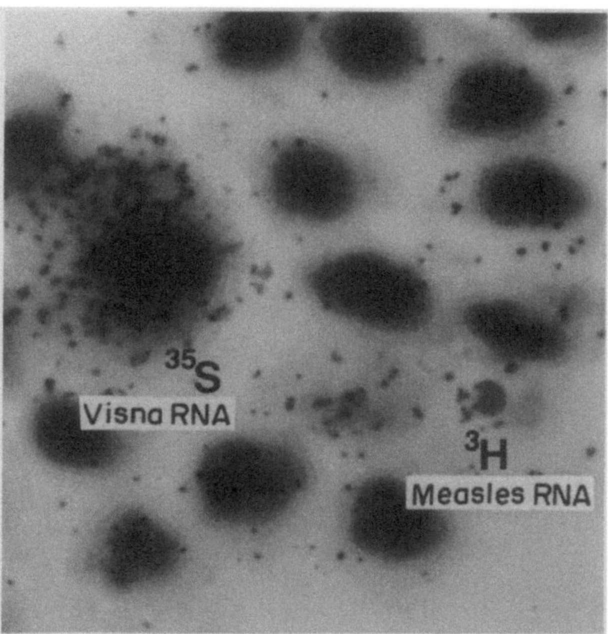

Fig. 6. Double-label in situ hybridization and color microradioautography. Mixed population of cells infected with visna virus or measles virus were hybridized in situ to probes labeled respectively with ^{35}S or ^{3}H. Grains in a first layer of emulsion have been converted to a magenta color, those in the second to a cyan color. Since only ^{35}S penetrates the second layer, the cell with visna virus RNA is identified by the superimposed magenta and cyan colored grains. (From HAASE et al. 1985, reproduced with permission.)

Figure on page 19

Fig. 6A–D. Colorimetric in situ hybridization in formalin-fixed paraffin-embedded tissue sections. ▶
Method modified from UNGER et al. 1986. **A** Cytomegalovirus in open lung biopsy from bone
marrow transplant patient. Biotinylated cloned cytomegalovirus probes detected with avidin-
alkaline phosphatase and McGadey reagent. Signal is dark brown/black. Hematoxylin counterstain.
B Adenovirus in lung from autopsy of child with severe combined immune deficiency following
bone marrow transplantation. Biotinylated whole genomic adenovirus 2 DNA detected as in **A**.
Hematoxylin counterstain. **C** Herpes simplex virus (HSV) detected in adrenal gland of a congenitally
infected infant. Biotinylated cloned HSV probe detected with avidin-alkaline phosphatase and
modified McGadey reagent (iodonitrotetrazolium substituted from nitroblue tetrazolium). Signal
is red/brown. Hematoxylin counterstain. **D** Human papilloma virus type 6 detected in cervical
biopsy of patient with persistent atypia on pap smear and coloposcopic abnormality. Biotinylated
cloned HPV 6 probe detected as in **A**. Neutral fast red counterstain

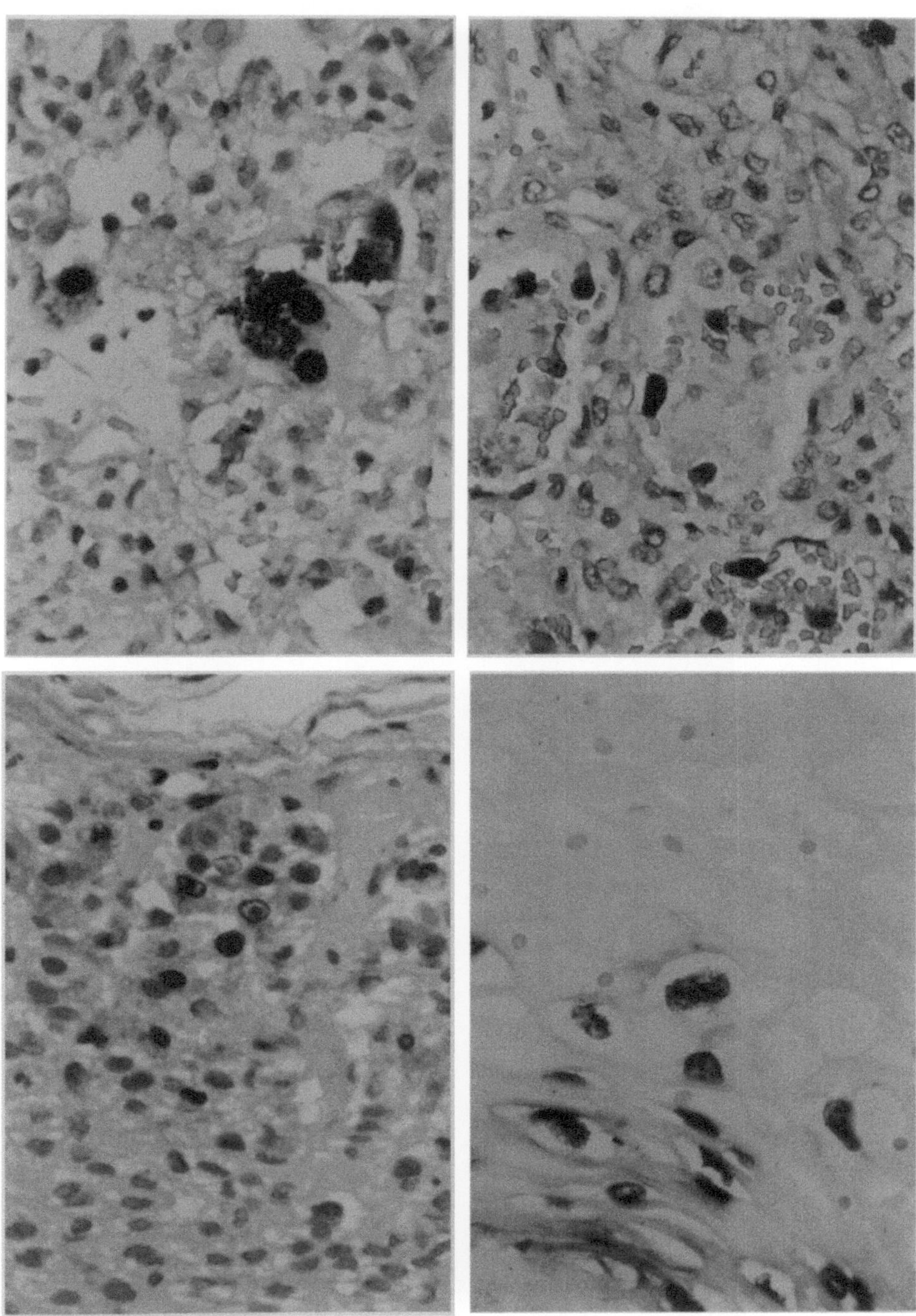

Figure on page 28

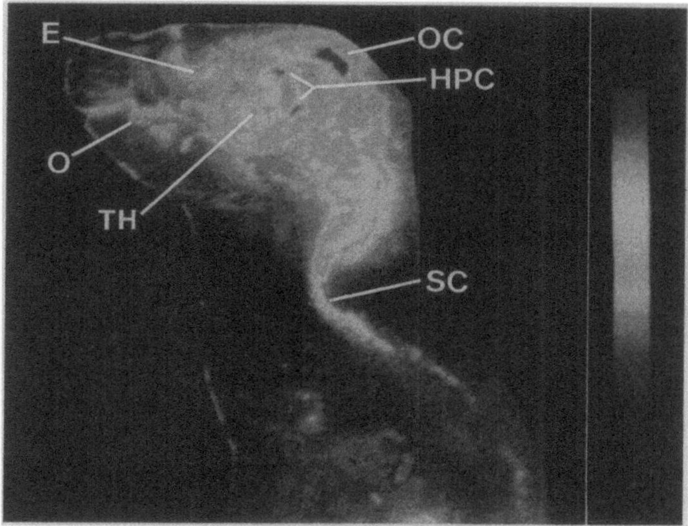

Fig. 5. Quantitative distribution of reovirus (REOV) antigens in a mouse acutely infected with REOV. Sagittal section incubated with rabbit antibody to REOV and ^{125}I-labeled staphylococcal protein A. Autoradiogram was scanned with a laser densitometer for color-coded densitometry. *E*, eye (retina); *HPC*, hippocampus; *O*, oropharynx; *OC*, occipitalcortex; *SC*, spinal cord. REOV antigen concentration is indicated by color: red is highest, yellow is intermediate, blue is lowest (color bar at *right* of figure)

Figure on page 39

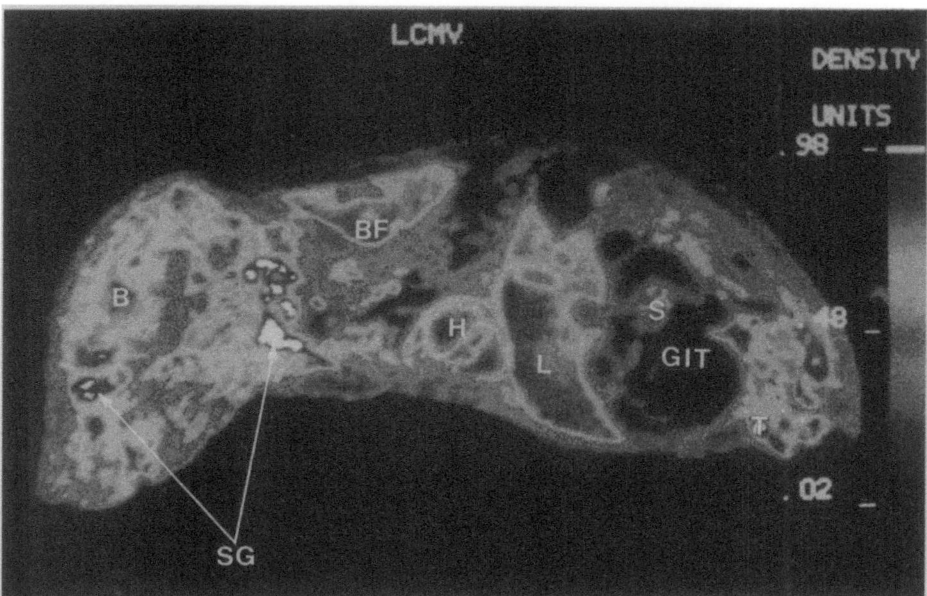

Fig. 6. Quantitative distribution of lymphocytic choriomeningitis virus (LCMV) antigens in a mouse persistently infected with LCMV. Sagittal mouse section incubated with guinea pig antibody to LCMV and ^{125}I-labeled staphylococcal protein A. Autoradiograph was scanned with a laser densitometer for color-coded densitometry. *B*, brain; *BF*, brown fat; *GIT*, gastrointestinal tract; *H*, heart; *L*, liver; *S*, spleen; *SG*, salivary glands; *T*, tests. LCMV antigen concentration is indicated by color: white is highest with decreasing concentrations shown in red, yellow, green and blue (color bar at *right* of figure)

Figure on page 40

MIX
Papier aus verantwortungsvollen Quellen
Paper from responsible sources
FSC® C105338

If you have any concerns about our products,
you can contact us on
ProductSafety@springernature.com

In case Publisher is established outside the EU,
the EU authorized representative is:
Springer Nature Customer Service Center GmbH
Europaplatz 3, 69115 Heidelberg, Germany

Printed by Libri Plureos GmbH
in Hamburg, Germany